ECOLOGY AND DISTRIBUTION OF RECENT FORAMINIFERA

Ecology and Distribution of Recent Foraminifera

BY FRED B PHLEGER

The Johns Hopkins Press: Baltimore

© 1960 by The Johns Hopkins Press, Baltimore 18, Md.

Distributed in Great Britain by Oxford University Press, London

Printed in the United States of America

    by Universal Lithographers, Baltimore

Library of Congress Catalog Card Number 60-11984

Second Printing 1965

Foreword

This book summarizes the principal work done on the ecology of modern Foraminifera and attempts to integrate the results into a useful system. It is mainly a statement of progress and a definition of a field of study. Most of the work on Foraminifera has been on their taxonomy and their use as stratigraphic markers for geologic correlation. The point of view expressed here is that Foraminifera are of value in problems of oceanography and paleoceanography.

It is hoped that the data and interpretations presented will be of value to both oceanographers and geologists. This does not necessarily mean that all of the suggestions which are made will find immediate application in those sciences. Many hypotheses are presented which are subject to change as new data and interpretations are obtained. It is hoped that the ideas presented may be useful in helping to direct certain studies in oceanography, sedimentology of ancient marine rocks, and in related fields.

An attempt has been made to include all important research on ecology of modern Foraminifera already published. Some references undoubtedly have been overlooked; such omission is inadvertent and no criticism of such work is implied. There are a few papers in the bibliography to which no special references are made in the text.

Grateful acknowledgement is made to the following people who acted as critics in reviewing the manuscript and in making numerous valuable suggestions: Ben B. Cox, Frances L. Parker, Roger R. Revelle, Philip C. Scruton and William R. Walton. Jean Peirson Hosmer assisted in editing and preparing the manuscript and illustrations. Much of the research of the Marine Foraminifera Laboratory has been supported by contracts of the Office of Naval Research with the University of California and with the Woods Hole Oceanographic Institution. Important support for our work also has come from the National Science Foundation, the American Petroleum Institute, and the Geological Society of America.

La Jolla, California
January, 1960

Fred B Phleger

Contents

Foreword   v

I. *Introduction and Methods*   1

Introduction   1
Some Characteristics of the Modern Ocean   4
Some Major Marine Environments   10
Methods of Study   19

II. *Depth Distribution of Benthonic Foraminifera*   39

General Statement   39
Depth Ranges of Species   45
Applications of Depth Range Information   90
Ecologic Factors Affecting Depth Distributions of Foraminifera   102

III. *Marginal Marine Distributions of Benthonic Foraminifera*   125

Distribution of Assemblages in Gulf of Mexico and Northeastern United States   126

Interpretation of Marginal Marine Faunal Patterns   136
Relationship between Faunal Patterns, Relative Runoff and Extent of Barrier Islands in the Northern Gulf of Mexico   145
Composition of Gulf Coast and Eastern North American Nearshore Faunas   148
Other Studies on Nearshore Distributions   161
Discussion of Marsh Faunas   175
Coral Reef Faunas   181

IV. *Distribution Studies of Living Benthonic Foraminifera*   186
General Discussion   186
Relative Rates of Deposition Based on Benthonic Foraminiferal Populations   189
Seasonal Production   204

V. *Studies of Planktonic Foraminifera*   213
Distribution of Living Planktonic Foraminifera   214
Distributions in the North Atlantic Based on Sediment Samples   220
Mixed Planktonic Faunas   225
Continental Shelf Distributions   239
Interpretation of Pre-Modern Planktonic Faunas in Submarine Cores   243

VI. *Summary Discussion*   254
Comments on Geologic Applications   259
Oceanographic Applications   265
Research Problems   270

References   277

Index   289

ECOLOGY AND DISTRIBUTION OF RECENT FORAMINIFERA

CHAPTER I

Introduction and Methods

Introduction

The study of ancient marine sediments is in large part concerned with the definition of ancient marine environments and their distribution through space and time. The definition of an ancient marine environment is based upon what is known, or presumed to be known, of processes and environments in the modern ocean. Only in the last few years has much specific information about marine distributions and processes been obtained which is useful in reconstructing ancient seas. The available oceanographic information has been only partially utilized by geologists in interpreting marine rocks.

The bottom of the ocean may be thought of as the dumping ground for: (1) debris washed in from the land, and (2) materials produced in the water. Abundant debris from organisms is produced, both as skeletal material and nitrogen-carbon compounds. The organically derived materials are being produced throughout the overlying water at or near the place of deposition. The materials accumulating on the bottom thus reflect the sum of the marine environment which affected them. The basic problem is to interpret the sedimentary materials occurring in ancient rocks in terms of a realistic ocean.

Skeletons of marine organisms are perhaps the most satisfactory materials for determining the environments in which marine sediments were deposited. Organisms exist in any marine habitat as a result of a complex series of physiological adjustments. They have specific adaptations to the environments in which they live, and many of them resist or do not tolerate radical changes in habitat. A knowledge of these adjustments, especially of the principles governing distributions of organisms, is essential to the understanding of conditions under which the detrital material in which organic remains occur was emplaced.

Description of marine sedimentary rocks and their enclosed fossils, correlation of strata, and delineation of the distribution of ancient seaways comprise only the first phase in understanding the history of the oceans of the past. To gain an understanding of the history of ancient oceans it is necessary to reconstruct their over-all physical, chemical, and biological characteristics insofar as possible. For example, if the surface and bathymetric distribution of water temperature, or temperature differences, can be determined, the marine air climate also can be inferred, including atmospheric circulation. This information, along with data on probable distribution of salinity, may be used to reconstruct the distribution, direction, and intensity of currents. Knowledge of the currents and other water masses will have an important bearing on the distribution of marine organisms and sediment. It may eventually be possible to reconstruct some of the bordering continental climates by knowledge of the effects of water masses on nearshore sediments. Marine chemical and biological distributions are a key to determining water-mass distributions and *vice versa*. It is theoretically possible to reconstruct many of the features of an ancient ocean in specific terms, provided we have the data and can interpret sedimentary rocks deposited in such an ocean.

The only apparently reliable point of reference is the study of both organic and inorganic sedimentary materials in the modern ocean. This reasoning is based on the assumption that the present ocean provides the key to past oceans. Such an assumption may be only partly true, but it is believed that the same kinds of physical, chemical, and biological processes occurred in ancient oceans as in the modern ones. There is only a general similarity between organisms of the Paleozoic and those of the present day. On the other hand, it seems probable that the same basic processes affected distributions of both types of organisms.

Any group of organisms which is abundant, diverse, and widespread both laterally and bathymetrically should provide an excellent tool for studies of paleoceanography. The Foraminifera combine all these attributes, and have the added advantage that they are well-known and are being actively studied by hundreds of workers for geological purposes. The ensuing chapters summarize most of the research which has been published on the ecology of modern Foraminifera and, in addition, include other results and interpretations presented here for the first time. An attempt is made to evaluate these data in terms of their probable usefulness in solving oceanographic and paleoceanographic problems. Suggestions are made for additional study where such effort promises to have significant results and applications.

Some Characteristics of the Modern Ocean

The science of oceanography is making substantial contributions to many basic problems in the interpretation of ancient marine rocks. If we can understand modern marine sedimentary environments, if we can interpret marine sedimentary and faunal facies, many of the problems of historical geology can be clarified. In studying marine sediments it should be remembered that the ocean is an environment in which all the factors are closely interrelated; therefore, such studies must be related to all the physical, chemical, and biological features of the entire water mass insofar as this is possible. Detailed sedimentary and faunal studies which are not correlated with known water characteristics are of only limited practical value. An understanding of the ecology of Foraminifera, as well as other marine organisms, must be based on a knowledge of some of the basic features of the ocean.

The following is a very brief description of a few of the major features of the modern ocean which are of value in interpreting environments under which modern and ancient marine sediments have accumulated. Additional and more detailed information may be obtained from Sverdrup *et al.* (1942) and also from such sources as *Collected Reprints of the Woods Hole Oceanographic Institution, Scripps Institution of Oceanography Contributions, Journal of Marine Research, Transactions of the American Geophysical Union, Deep-Sea Research,* and various oceanographic expedition reports.

MAJOR PHYSICAL AND CHEMICAL ASPECTS

One of the most obvious features of sea water is that temperature varies from place to place and time to time. Variations of surface water temperature with latitude in the North

Atlantic are shown in Figure 1; these are taken from charts of mean monthly surface temperatures compiled by the U. S. Navy Hydrographic Office. The mean February temperatures are the *minimum* mean monthly temperatures in the northern hemisphere; the mean August temperatures are the *maximum* mean monthly temperatures in the northern hemisphere. It should be realized that these charts of mean monthly temperatures are generalizations and are thus subject to various qualifications. The actual surface temperatures existing at any time in the North Atlantic only approximate the distributions shown on these charts, and actual maximum and minimum temperatures are somewhat different.

In addition to variation with latitude, there is also a vertical thermal structure. In mid-latitudes the winter water is relatively cold and well-mixed to a depth as great as 100 m. or more. This turbulent and cooled water may thus cover the entire continental shelf. By summer in mid-latitudes the water has warmed up and a seasonal thermocline usually develops in the upper few hundred feet. These are the two extreme stages in the yearly temperature cycle of the upper water. The water showing seasonal temperature variations (in mid-latitudes) is termed the seasonal layer (see Figure 2a).

In many areas of the tropics this upper layer of water is always warm, reflecting the surface temperature, and generally well-mixed and isothermal to a depth as great as 150-200 m. in the Trade Winds belt (Figure 2b). Seasonal temperature variations within the seasonal layer may be insignificant in such an area.

Below the seasonal layer is the permanent thermocline in which there is a decrease of temperature with increase in depth. The temperatures in this layer, although somewhat variable, remain relatively constant with depth in any region. The permanent thermocline extends to an average depth of approximately 1000 m. The deep, bottom water is at depths

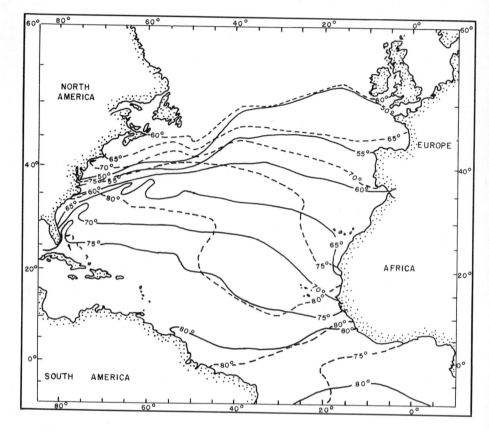

Figure 1. Maximum (August) and minimum (February) mean monthly sea surface temperatures in degrees Fahrenheit in the North Atlantic. February, solid line; August, dashed line. From Phleger, Parker and Peirson (1953).

CHAPTER I 7

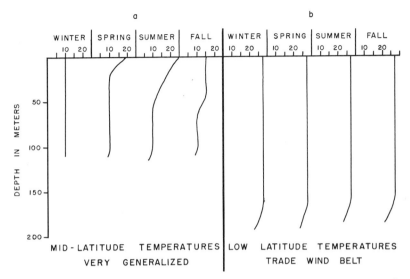

Figure 2. Generalized seasonal temperature gradients in mid-latitudes and the tropics.

greater than about 1000 m. and is uniformly cold. Generalized bathymetric distributions of marine temperatures are illustrated on Figure 41 in Chapter II.

Temperature is important in affecting the distribution of organisms and oceanic circulation. It is apparent from this brief summary that: (1) surface-water temperatures do not necessarily define the thermal characteristics of any body of water; (2) isolated and/or single temperature measurements do not define the temperature characteristics in the seasonal layer, especially in mid-latitudes; (3) single observations of bottom temperature are of little or no significance in mid-latitude shallow water; (4) there are three distinct bathymetric water layers: the seasonal layer, the permanent thermocline, and the deep bottom water; and (5) general inferences about temperature conditions may be made by understanding temperature distributions in a few critical areas.

Ocean water is a complex chemical solution in which there are many reactions and cycles not yet fully understood. Offshore sea water, away from influences of land, has approximate surface salinities of 33-37 o/oo, is alkaline (pH 8.0 ±), and is a strongly-buffered solution. In the open ocean the ratios between the most abundant dissolved materials remain essentially constant. Many materials which are of critical importance in biological processes have variable distributions seasonally and laterally. These include oxygen, carbon dioxide, nitrate, phosphate, silica, etc. Various elements present in very small amounts also may have variable concentrations; some of these trace elements are thought to be limiting factors in the distribution of certain organisms.

In nearshore areas where land runoff is high and salinity is lower than in offshore areas, ionic ratios no longer are constant. It is apparent, also, that other chemical characteristics of sea water are modified under these conditions, such as pH, trace element composition, etc.

The major surface circulation in the ocean is well-known. It is a result principally of wind stress on the water surface, and is modified by differences in water density due to variations in temperature and salinity, by the rotation of the earth, and by the distribution of the land. In general, it may be said that the surface currents reflect the wind circulation and are modified by other processes and factors. Recent study of the Gulf Stream, the best-known major ocean current, has shown that in many places it is narrower and swifter than formerly believed; it also meanders and has short-period lateral variations in position up to several miles (Fuglister and Worthington, 1951). Significant tidal currents are often developed in nearshore areas when there is sufficient lunar or wind tide and the configuration of land or other factors tend to inhibit the free flow of water.

Currents are of considerable importance in marine sedimentation in a variety of ways: (1) they act as a transporting agent for inorganic sedimentary particles, (2) they retard or prevent sedimentation in areas where they directly influence the bottom, and (3) they distribute planktonic organisms, which may become important constituents of the sediment, and larval stages of various benthonic organisms.

Current convergence and divergence areas are of importance, especially in affecting the distribution of planktonic organisms. Convergence occurs where surface water from different sources comes together and either sinks or mixes on the surface. Examples of convergence are: (1) The Antarctic and sub-Arctic convergences where there is a mixing of mid-latitude and high-latitude water, and (2) the convergence of equatorial counter-current water with mid-latitude water to form the equatorial current on the eastern sides of the oceans. Divergence occurs where surface currents split or where upwelling replaces surface water which has been displaced. In convergence two different types of water from different sources are mixed, and in divergence there is spreading of water from one source.

DISTRIBUTION OF ORGANISMS

Vast amounts of organic material are produced in the ocean. All the primary organic production is confined to the upper water layers where light penetration is sufficient to permit photosynthesis. Most photosynthesis occurs at water depths of less than approximately 50 m. although in tropical offshore areas it may occur as deep as a few hundred meters. Among the principal producers are algae, such as diatoms and dinoflagellates and similar forms which are not well known because of their small size. H. Moore has stated (personal communication) that in his experience these forms which are too small to be caught in a net appear to constitute the major part of the plankton population. The production of organic materials is dependent upon the supply of plant nutrients in the upper water layers. Replacement of nutrients to upper water layers is largely from nutrient-rich deeper water. The deeper water from a few to several hundred meters depth is brought to the surface by wind mixing and surface divergence.

Different assemblages of organisms are adapted to different kinds of marine water. Examples of such water bodies are: lagoon water, continental shelf water, deep-sea water, Gulf Stream water, Arctic water, etc. Organisms (both planktonic and benthonic) may be used to identify and trace boundaries and movements of these waters. The factors of the environment which limit distributions of faunas are either unknown or very poorly known. It may be said, however, that each type of water containing a distinctive fauna and flora seems to have a specific and uniform ecologic effect, although the factors causing such distributions may be unknown. This concept is of value in describing and interpreting faunal patterns.

Planktonic organisms, especially zooplankton, live throughout the entire water column and may be affected by environmental differences at various depths. Phytoplankton are restricted to near-surface waters. Benthonic organisms are affected by the nature of the bottom water and directly and indirectly by the sum of environmental features at and near the bottom on which they live.

Some Major Marine Environments

It is of interest to examine some different marine habitats and to define their characteristics as environments for organisms. No attempt is made in this brief discussion to make a complete analysis of all marine and near-marine environments described in the literature. Although much has been written on this subject, information still is quite incomplete. The marine environments given below are those which have been found useful in studies of foraminiferal ecology; the descriptions are generalized. The classification illustrates many of the principal features of marine environments.

MARINE MARSH

The most obvious component of the average marine marsh is the large quantity of plant material, such as salt grasses and mangrove thickets in tropical areas. Another important feature is the great variation in type of plant material over short distances; this is especially notable in the salt-grass type. Many marsh areas support a large population of animal as well as plant life. Large populations of Foraminifera often are found in this habitat, and there is great variation in population density from place to place, with large populations separated by only a few feet from places where there are few or no living Foraminifera. The reasons for this are not clear, but it appears to be related to variations in environmental conditions.

The water in most marshes is shallow, seldom deeper than a few inches, although tidal channels may be several feet in depth. Where there is a tidal range of several feet, most of a marsh may be subjected to complete withdrawal of water. In some marshes there are "high marsh" areas which are wetted only from beneath except during times of very high tide.

Salinities of marsh water vary depending upon the local fresh-water runoff, but are generally lower than "oceanic" salinities. In areas where runoff is negligible much or most of the time, the marsh water has a salinity comparable to or greater than that of the nearshore marine water nearby. In marshes which are subjected to occasional flooding by marine and fresh water or heavy local rain, great variations in salinity are to be expected. In general, salinity decreases inland from the shore, and may vary considerably from one pool to the next. Some marshes, however, are hypersaline at least in part. The writer has measured salinities in marine marshes occurring in a desert area which are as high as 48 o/oo, considerably greater than normal sea water.

A great range in temperature may also occur. In many midlatitude areas water temperatures probably vary from about 0°C. to 35°C., especially where a small body of water is isolated from its source of supply. In the tropics there may be also a considerable diurnal temperature range, especially in exposed areas.

ESTUARY

The term "estuary" is used here in a rather general sense for a river which is subject to invasion of marine water for some distance landward from the mouth of the river. The characteristic of the salt water in such an estuary is that it usually is considerably more dilute than the marine water, due to mixing with the fresh water of the river. Salinities may vary from only a trace to those of the open sea. Generally, fresh water floats on top of a "salt-water wedge" which invades along the bottom of the river, as shown in Figure 3. Such invasions of salt water along the bottom may occur for 50 miles or more, depending upon the slope and depth of the

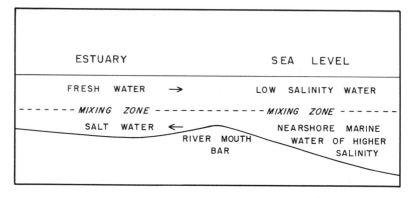

Figure 3. Schematic cross-section of an estuary and the nearshore area near the mouth of an estuary.

river bottom. Salinity of such bottom water decreases upstream. The salt-water wedge oscillates with the tide and is modified by runoff. Occasionally, the entire cross-section of the river may be fresh, when during a time of flood the salt-water invasion is prevented by hydraulic pressure of surplus fresh water. When there is little or no runoff the channel may carry essentially marine water.

The range in water temperature in such an environment will depend upon its geographic position, the climate of the region, and the amount of runoff. Considerable range occurs, especially in mid-latitudes.

The estuary is intimately associated with the surrounding marsh which is usually present, at least near the mouth of the estuary, and the water from one flows into the other at frequent intervals. Where an estuary is very narrow it is frequently difficult to distinguish any marked boundary with surrounding marsh. The river water may be a source of abundant food in the form of plant material and certain dissolved nutrients. This is suggested by the unusually large living populations of Foraminifera off the mouths of the Mississippi and the Guadalupe rivers. The proportions of the dissolved salts are different from those in normal marine water, and river runoff furnishes certain salts which may be less common in marine water. Water turbulence in such a habitat is expected to be relatively intense and to show considerable variation with position and time. Many estuaries are very turbid.

COASTAL LAGOON

A coastal lagoon is a body of water which is partially enclosed by a land barrier and which has restricted communication with the sea through one or more inlets. Many lagoons are only a few feet in depth, but some may be 100 ft. (30 m.) or more deep.

From an ecological point of view coastal lagoons are of different types, depending upon the climate and the runoff from the adjacent land. A lagoon on the margin of a high-runoff area will contain water of lower salinity than that of adjacent sea water, often in the range of 20 o/oo to 30 o/oo. If the runoff is consistently high the abundance of fresh water will tend to form a dynamic barrier to the invasion of undiluted marine water over a wide front. Marine water will invade, however, along the bottom and will be mixed quickly with fresher water if the depths are shallow.

If runoff is variable and ranges from very high to very low, lagoon salinities will vary from very low to very high. As shown in Chapter III (see Figure 51), in this environment the geographic barrier (island or spit) tends to be continuous except for an occasional inlet. In a dry climate the condition most of the time is that of very low runoff, and the occasional rains usually cause flood conditions. The supply of fresh water in great abundance during floods rapidly lowers the salinity in the lagoon and prevents the invasion of much marine water, especially if the lagoon is shallow—the usual condition. Salinities under such circumstances may fall to less than 10 o/oo and persist as long as flood conditions persist. In the long, dry periods between floods all the water in the lagoon is essentially marine in origin and may attain salinities up to 45 o/oo or more due to evaporation of marine water.

If there is little or no runoff the water of the lagoon is furnished entirely from a marine source except for rainfall on the water. If the climate is dry, evaporation will cause concentration of the marine salts, and salinities of 100 o/oo or more may result in extreme cases. Two examples of this type are Laguna Madre in the Gulf of Mexico on the south coast of Texas (Rusnak, 1960) and Laguna Ojo de Liebre on the Pacific coast of Baja California, Mexico (Phleger and Ewing, in press). Scruton (1953a) has given an excellent theoretical discussion of high-salinity lagoons.

The food supply in a lagoon may be relatively abundant, especially near a river, and the organic production may be considerable. The writer has found unusually high populations of living Foraminifera in San Antonio Bay, Texas, near the mouth of the Guadalupe River, and also off the mouth of Pass a Loutre in the Mississippi Delta area.

In mid-latitudes there may be a considerable range in water temperatures in lagoons, especially in shallow ones. Known temperature ranges in lagoons along the northern coast of the Gulf of Mexico are from less than 40°F. (4°C.) to more than 90°F. (32°C.). Even in the tropics there may be a large diurnal temperature range in shallow lagoons. Considerable turbidity is expected because of suspended sediment load introduced by streams or put into suspension by wave action in shallow water. Turbulence is high, but less than in the surf zone of an exposed beach. Rapid and great changes of salinity and temperature in many lagoons in the mid-latitudes may cause widespread mortality among the organisms living there.

BEACH AND NEARSHORE

The beach and nearshore environment includes the area along exposed coasts where there is surf action. The principal characteristic of this environment is extreme water turbulence. As a result the bottom usually is composed of sand, where detrital sediment is available, and there is much shifting of the substrate. Marked turbulence occurs at all times to a depth of 30 ft. (9 m.) at La Jolla, California, according to D. L. Inman (personal communication) and is observed frequently at 100 ft. (30 m.) when long-period waves arrive. Temperature, salinity, and other water characteristics of this environment are those of the nearshore surface water and will vary from place to place.

Organisms which grow in the beach and nearshore environment are those which can withstand the rigors of water turbulence and shifting substrate. There are large numbers of sand-burrowing organisms of the hardy type and forms with heavy

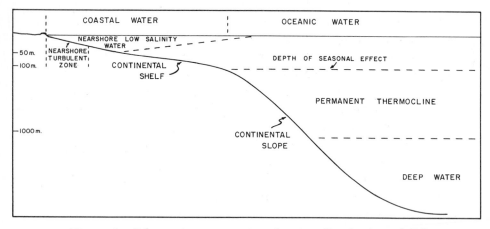

Figure 4. Schematic cross-section showing distribution of different types of water in the open ocean offshore from an area of high runoff.

shells. Considerable destruction of shell material undoubtedly occurs. Also, the fact that the sand is well-sorted means that much skeletal material smaller than sand will be transported out of this environment. As shown in Chapter III, only the large, thick-shelled Foraminifera escape transportation and/or destruction of their tests. Diatom frustules, for example, generally are not found in beach and nearshore sediments, probably because they are too small. Rocky areas in the nearshore zone have a fauna which differs from that in sandy areas, especially in the presence of numerous attached forms.

CONTINENTAL SHELF

Coastal water, as distinguished from offshore, "oceanic" type water, extends approximately to the edge of the continental shelf in many places, as shown in Figure 4. The depth of the seaward edge of the continental shelf is approximately 100-125 m. (55 fm. ±) on the average, and the distance from shore may vary from less than a mile as off parts of California, up to a hundred miles or more, as off the coast of Florida. The water above the shelf, in a very general way, represents a distinctive environment which may be quite different from offshore water.

One of the principal characteristics of the shelf water is that it is mostly or entirely within the seasonal layer, and the seasonal effect extends to the bottom (Figure 4). In the tropics, where there is little or no seasonal variation, the zone of mixed isothermal water frequently extends to this depth or deeper (Figure 2). The water within the seasonal layer has considerable variation in the mid-latitudes, and it is also the zone of most organic production. It is more turbid and turbulent than deeper water layers. This environment may be characterized as being more variable than deeper environments. There is an apparent general gradation of sediments from coarse material nearshore to progressively finer sediments offshore, if the present cycle of deposition only is considered.

Along coasts having moderate to high runoff the upper water may be of lowered salinity at least as far as the edge of the shelf, and this may go as deep as 180 ft. (55 m.) or more. In the northern Gulf of Mexico the benthonic foraminiferal faunas of the inner shelf are different from those of the outer shelf (Phleger, 1956) with a boundary at approximately 150-180 ft. (46-55 m.). It is presumed that on the inner shelf influences of runoff and deposition from the land are more pronounced and on the outer shelf that "oceanic" influences are dominant (see Figure 4). The environment on the inner shelf is expected to be more variable than that on the outer shelf.

CONTINENTAL SLOPE AND DEEP-SEA

The continental slope and deep-sea include a series of related environments which are somewhat different, but may be grouped together for convenience and because they are closely related. The upper 800-1000 m. are within the permanent thermocline, so that there are different temperatures at different depths except in high latitudes where the temperature is uniformly cold at all depths. The depth of 800-1000 m. may be used as a boundary between the upper and the lower continental slope. A microfaunal boundary is recognizable at this depth in many parts of the world. Below the permanent thermocline temperatures are generally 5°C. or less. The salinity of most of this oceanic water is about 33 o/oo to 36.5 o/oo, except in areas like the Mediterranean Sea where the salinity is somewhat higher due to excessive surface evaporation and low runoff. The organic production is confined approximately to the upper 300 m. overlying this province, even in the clear, oceanic water of the tropics. The substrate is composed of soft mud. There is little turbidity or turbulence on the bottom at most times and places, and in general this environment differs from other marine habitats in being more uniform. Photographs of ripple marks at great depths show the existence of deep-sea currents, and there may be relatively little or no sedimentation on topographic highs such as seamounts. Most of the ocean bottom is in this generalized province.

Methods of Study

Foraminifera are single-celled animals, or Protozoa, most of which are marine although a few forms have been reported from fresh water. Most forms have a test composed of calcium carbonate, secreted by the organism, but tests of agglutinated sand grains or other materials are common. A few are composed of chitin and many species have a chitinous inner lining in an otherwise agglutinated test.

Modern Foraminifera live in all marine and semi-marine environments and are both planktonic and benthonic. There is a great number of species and much variation in individuals of some species. Any population of these organisms should reflect all the environmental factors which exist where they occur. It is necessary, therefore, to understand as much as possible about the environmental relationships of these forms and the reasons for such adjustments.

The geological importance of the Foraminifera is that they produce a shell or test. When the animals die or reproduce, the tests are added to the debris accumulating on the bottom. They thus become an integral part of the sediment and leave a record of conditions and events in the ocean above and around the sediment. In addition, they behave as grains of sediment and may be transported and eroded by marine processes which cause transport within their size range.

There is a voluminous literature on both fossil and modern Foraminifera. Most of the work is of two general types: (1) taxonomic studies which define the various species, etc., and attempt to show their evolutionary relationships by biological classification, and (2) stratigraphic studies in which these forms are used to determine the geologic ages and age correlations of stratified rocks. There are excellent summaries of these studies by Cushman (1948), Galloway (1933), Glaessner (1945), Sigal (1952) and others. The present discussion is confined to ecology of modern and near-modern Foraminifera.

Study of the ecology of modern Foraminifera has been approached from various points of view and by different methods, and it is not the present purpose to summarize all the methods which have been used. Many of the results presented are from studies by associates of the Marine Foraminifera Laboratory, and the methods described here are those in current use. Ecology of Foraminifera has been studied mainly through distributions of species and faunas, and by correlations of those distributions with what is known or inferred about the marine environments in which they occur.

Physiological experimentation and observation of living specimens, desirable parts of any ecological study, have been carried on by a number of workers including Myers (1935a, b, 1936, 1937, 1942, 1943), Le Calvez (1938), Arnold (1954a, b), and at the Marine Foraminifera Laboratory by Bradshaw (1955, 1957). There are two apparent inherent difficulties in experimental work on the Foraminifera for ecological purposes:

1. There are numerous modern species, all of which probably have different adjustments to factors of the environments in which they occur. There is thus no reason to expect that results obtained on one species will apply to any other species. Since physiological research is laborious and difficult, a considerable amount of time and effort are required for reliable results.

2. It is not possible to reproduce adequately the marine environment in the laboratory.

FIELD METHODS

Remains of Foraminifera are so abundant in marine sediments that they are well-suited to quantitative, distributional studies from relatively small samples. Marine sediment samples for reliable population studies of modern Foraminifera should be collected with considerable care and should have the following characteristics: (1) known surface area, (2) known volume of sediment, and (3) undisturbed material from the actual sediment surface. Samples which do not satisfy these requirements, insofar as practicable, are not considered to be reliable for population studies.

Sediment Sampling. Basic methods of collection to fulfill these requirements are now well-established, although modifications are made with continuing experience. It seems desirable to describe them in some detail here so they may be used by other workers. The standard collecting instrument for shipboard use is a small coring tube (Phleger, 1951b). This sampler (Figure 5) consists of four parts: a steel or brass outer tube, a plastic inner liner, a water valve at the top of the tube, and a lead weight. The steel or brass tube has an outer diameter of 1⅝ in., and an inner diameter of 1½ in. Approximately 30 lbs. of molded lead are fastened to the outside of the steel tube to force the sampler into the sediment. The lower end of the steel tube is fitted with a brass sleeve containing a threaded connection so that the tube below the weight can be detached, either for replacement in the event of damage or to remove the plastic liner. There is a hardened-steel cutting edge which fits on the lower end of the tube by a bayonet-type lock.

The sampler weighs 35 to 40 lbs. and can be operated from a light-weight, high-speed winch of the type used for bathythermograph observations. The depth of successful operation with this type of winch probably does not exceed about 500 m. In deeper water the coring tube may be used with the heavier and slower hydrographic winch which is usual equipment on an oceanographic research vessel. The sampler is allowed to plunge into the bottom sediment by its own momentum. During operations in deeper water a tripping device similar to that described by Hvorslev and Stetson

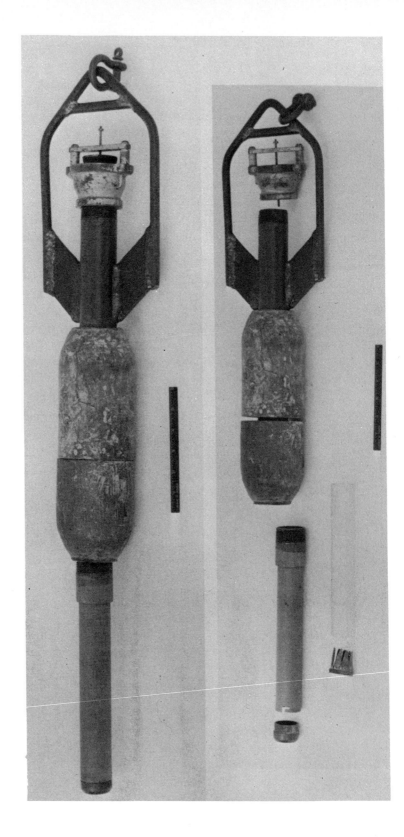

(1946), but smaller in size, has been used. This tripping device is seldom necessary since satisfactory cores are obtained without its use.

In very deep water (deeper than approximately 1000 m.) it is often impossible to ascertain when an instrument such as this reaches bottom because of the light weight of the sampler relative to the weight of the wire. This problem can be obviated by use of the "ball-breaker" assembly described by Isaacs and Maxwell (1952), in which the sound of an imploding glass ball can be recorded on the deck of a ship.

In deep water the small coring tube frequently is used as a pilot weight on the tripping device of a large coring tube, is attached to the frame of a submarine camera, or may be a pilot weight on the wire being used for hydrographic observations. It is well adapted for these purposes, and obtaining a sample in such a manner saves time and thus expense.

The purpose of this small coring tube is to obtain a relatively undisturbed sample of known area of surface sediment. It is important that the tube be handled properly to obtain a satisfactory sample. The operating instructions which follow are the result of experience in collecting several thousand samples and in analyzing their quality by laboratory study.

The coring tube should be carefully washed free of all sediment before use to avoid contamination between faunas of Foraminifera or other materials. The corer should be pulled out of the bottom material as gently as conditions will permit. When it reaches the ship's deck the sampler must be kept upright. A cork is placed in the lower end of the tube and the plastic tube containing the sample is extracted by unscrewing the lower end of the steel tube, care being taken

---

Figure 5, Opposite. A small coring tube designed to obtain a relatively undisturbed surface sediment sample.

to keep the core upright. The water above the surface of the core is presumed to be the water which was directly above the bottom at that place and may be used for any desired analyses. To obtain the surface sample for living and dead Foraminifera, this water is carefully siphoned or poured off to within about one inch of the sediment surface. The remaining water is poured into a small wide-mouth jar. The core is extruded at the upper end until the sediment fills a measuring cutter (Figure 6) which holds the upper 1 cm. of the core. The upper 1 cm. is cut and placed in the small jar containing the water from directly above the sediment. Neutralized formalin in a concentration of approximately 5-10% is added. One of the most satisfactory materials for neutralizing formalin is hexamethylenetetramine (hexamine). Two pounds of this material added to one gallon of formalin will produce a mildly alkaline mixture. Additional neutralizing agent is added, usually sodium carbonate, and the sample is stored for future study.

This small coring tube will retain sediments containing sufficient silt or clay to give them cohesiveness. It will not retain most sands and other collecting methods must be used for such sediment. One of the samplers now in use for obtaining shallow-water sand (to 150 m.) is the modified dwarf orange-peel dredge (Phleger, 1952b) illustrated in Figure 7. This instrument can be used to collect reliable samples for living and dead population studies if the instructions listed below are followed. Walton (1955) shows that such samples are as reliable for studies of populations of living Foraminifera as those collected with the small coring tube.

The orange-peel dredge should be thoroughly washed free of sediment before use to avoid contamination of faunas. After the dredge is filled it should be brought to the surface slowly and with as little jerking as possible. If sea conditions permit, the sampler should be allowed to drain slowly at the sea surface before being brought on deck. It is then

Figure 6. Cutting device for taking upper 1 cm. of small core.

Figure 7. Dwarf orange-peel dredge, modified for sampling marine sediment.

held upright on deck with the jaws closed. The drawstring at the top of the canvas hood is loosened and a sample is collected from the center of the sediment in the dredge by inserting a short piece of plastic tubing of the size used in the short coring tube and corking it at the top. The surface sediment in the plastic tube is sampled for study of Foraminifera in the same manner as that described above for the corer. It is necessary to add sea water (not fresh water) to a sample collected in such a manner. Another sampler now used is a small model of a van Veen grab (Hedgpeth, 1957, Fig. 4 (2a)).

In very shallow water it is possible to collect reliable samples by pushing the plastic tubing into the sediment by hand. An extension may be made of steel tubing or a wooden pole which can be used in water as deep as 10 ft. or more. After collection, these samples should be treated in the same manner as described above.

Population studies of modern Foraminifera have been made on bottom sediment samples collected by several methods other than those described. One of the common bottom samplers is the snapper type. This instrument obtains a variable and unknown area of sediment and may sample considerably below the surface, especially in soft sediment. Dredges which are towed along the bottom obtain only a general mixture of sediment from a variety of depths within the sediment, and not necessarily a representative sample of the materials present. Occasionally foraminiferal studies have been made on samples taken from the flukes of ships' anchors. This type of sample has the following disadvantages: (1) only certain cohesive sediment will stick to the anchor, (2) the anchor flukes dig into the sediment for an unknown distance, and (3) there is a good probability of mixing between samples since anchors are seldom completely clean of sediment. Samples collected by these methods are of limited value in foraminiferal studies, are not reliable for adequate population analyses, and may give misleading results.

Long Cores. Results on studies of foraminiferal faunal sequences from long submarine cores are reported in a later section. A few comments about instruments for collecting such materials, however, may be useful. The simplest type at present in use is the gravity coring tube. The best-designed tube of this type is the Hvorslev-Stetson (1946) corer (Figure 8); this sampler falls free for several feet through the action of a tripper and obtains a core which is emplaced in a plastic or brass liner tube. A gravity coring tube seldom obtains a sample greater than 10 or 12 ft. (3-4 m.) in length, although one core 17 ft. (5 m.) long is reported by F. P. Shepard (personal communication). The piston coring tube (Figure 9), first developed for use in submarine geology by Kullenberg (1947), is now widely used. This instrument has obtained cores more than 25 m. in length; other models have obtained shorter cores, but generally longer ones than those taken by other methods. An excellent modification of the piston corer is described by Emery and Broussard (1954). In most of the piston coring tubes the sample is emplaced in an inner liner tube. In long cores, the surface sediment sample may not be reliable for sedimentary or faunal studies. This is due to drag of the surface sediment along the sides of the tube and mixing during handling of long cores in a horizontal position. It is now general practice to supplement a long core with a short core taken at the same time to obtain the surface layers.

Plankton Tows. Collection of living planktonic Foraminifera is essential to the understanding of their distribution. It is desirable to obtain a quantitative sample of plankton strained from a known volume of sea water. The Clarke-Bumpus (1950) plankton sampler (Figure 10) is designed to open and close at approximately known depths and the approximate amount of water strained is measured by counting the revolutions of a metering wheel. The small tow net is made of no. 8-19 silk bolting net with an average opening of 0.2-0.076 mm.

Figure 8, Opposite. Hvorslev-Stetson gravity coring tube.

Figure 9. Piston coring tube designed by Kullenberg.

These samplers operate satisfactorily at shallow depths but considerable difficulty often is experienced in operating them at depths greater than 200 or 300 m. The sampler should be modified by a stiffening rod extending from the sampler frame to the rear end of the net, as shown in Figure 10. The catching bottle is tied to the rod which prevents the net from becoming tangled with the towing wire and consequently damaged. This sampler should be towed at a speed of approximately 1½ to 2 knots.

Planktonic Foraminifera have been collected from numerous stations in the Pacific by vertical hauls from 50 m. with a small open-mouth net, shown in Figure 11. The volume of water strained by this method is calculated by the vertical distance towed, assuming that the entire cross-section of water traversed is strained by the net. The Clarke-Bumpus sampler is calibrated by the same method. This assumption introduces an error into the volumetric measurements, but the error probably is consistent.

LABORATORY STUDY

The methods of study which are briefly summarized are those used at the Marine Foraminifera Laboratory and are the result of experience in analyzing several thousand samples. These methods are constantly modified. It has been our general practice to deal with large numbers of samples instead of analyzing only a few in great detail, and it is believed that the advantages of this approach have been greater than the obvious disadvantages. On the other hand, more intensive study of single samples sometimes is desirable for specific purposes,

---

Figure 10, Opposite.   Clarke-Bumpus quantitative plankton sampler.

Figure 11, Far Right.   Plankton tow net used for vertical or horizontal towing.

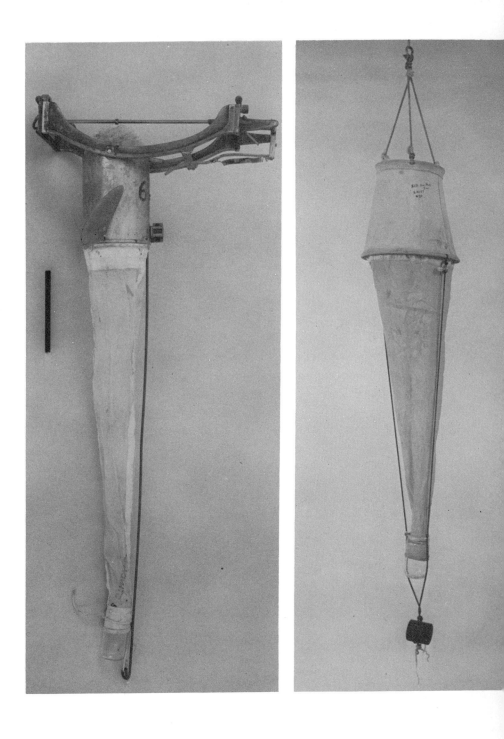

and this becomes more necessary as our knowledge increases.

Identifying Living Populations. The study of living populations as well as dead populations is essential in understanding the ecology of modern Foraminifera. In all methods used recently it has been assumed that a specimen in a sample preserved in formalin was alive at the time of collection if it contained protoplasm. This seems to be a reasonable assumption since decomposition in the sea is rapid and may be assumed to be almost instantaneous. Recognition of the presence of the preserved cell in a specimen without use of a color aid, such as a biologic stain, is laborious and not well-suited to rapid examination of large suites of specimens.

The most positive method for identifying "living" Foraminifera is Walton's (1952) staining technique with rose Bengal. A solution is made by adding enough rose Bengal to water to produce a dark red color. More or less stain may be used to obtain differing intensities of color, but it is probably desirable to use a solution of consistent concentration to obtain uniform results. Stain is added to the preserved sediment sample and this mixture is allowed to sit for several hours or, preferably, several days. The sediment is then washed over a sieve having openings of 0.062 mm. (or any desired size) and is placed in a clear plastic counting tray. Protoplasm may be recognized by its red or deep rose color.

Rose Bengal will give a deep rose color only to protoplasm, and in general will not greatly affect mineral matter, but occasionally a chitinous test may be slightly stained. Care must be exercised in distinguishing foraminiferal protoplasm in a test from other protoplasm which has been introduced from a foreign source. Many tests contain irregular, small particles which are stained; most of these are believed to be clusters of bacteria or other foreign micro-organisms. Small marine worms are common in many specimens, and these can be recognized by their shape. The foraminiferal protoplasm usually is clearly shaped to fit the test, or occasionally is concentrated in a globule in one of the chambers. Frequently only the protoplasm in one chamber or even a part of a chamber is stained, regardless of the length of staining time. This has no apparent significance other than reflecting the permeability of the specimen to penetration of the stain.

Study of Total Populations. After the living population has been counted wet, the sample is dried and only a fraction of the total population is counted in a large sample. One sample splitter described by F. L. Parker (1948) consists of a V-shaped metal trough bisected by a knife-edged partition. This splitter may not give so accurate a subdivision as the Otto Microsplit (Otto, 1933) which also has been used.

At least 300 specimens are identified and counted in samples having a population larger than that, and in actual practice the number counted usually varies from approximately 300 to 500 specimens. This is based in part on experience in dealing with many samples and in part on an analysis by Dryden (1931) of the accuracy in percentage representation of heavy mineral frequencies, a comparable problem. Dryden shows the relationship between the frequency of a member of a population and the accuracy of its frequency determination in any size population. The accuracy obtained by counting a fraction of a population larger than 300 individuals increases at a very slow rate. Figure 12, showing some of these probable errors, is reproduced here as an aid in evaluating such frequency data on populations of Foraminifera.

The estimate of the total populations depends in part on the size of the fraction counted. If the whole sample is counted, the total population is known within a small error, but if only 1/512 sample is counted an error of only one in the counted portion will produce an error of 512 in the estimate of the total. Although the absolute error varies with the fraction counted, the same percentage error will be present in the counted portion and the estimated total. We may speculate that an error of about 10% exists in the actual counting and

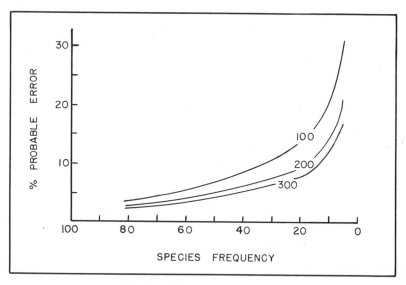

Figure 12. Probable error of a given frequency, counting 100, 200 and 300 specimens. Modified after Dryden (1931).

identification and an entire sample containing 200 specimens will give a total population that is accurate to 20. A 1/512 portion containing 200 specimens will give an estimate of the total population that is accurate to 10,000. There is also an unknown error in quartering of the sample to manageable size. The error in frequency of a species in a sample depends not only on the total number of specimens counted but also on the individual frequency. The error in the frequency of an individual species decreases as that frequency approaches 100% (see Figure 12, which gives the probable error for 100, 200 and 300 specimens counted).

It is assumed that any sample is representative of a large area of sediment, depending upon the spacing of the samples. In some instances this can be demonstrated to be approximately true since the assemblage frequencies can be plotted in a reasonable and consistent pattern (see Phleger, 1952b). It is true, nevertheless, that the populations are different in any two samples, just as any two specimens of the same species are different. There is a personal error in identification of species which cannot be completely eliminated. Other less important ones are possible in handling and processing the sample.

These considerations indicate that little if anything is to be gained by counting samples much larger than approximately 300 specimens and that the illusion of accuracy tends to be misleading. These qualifications in population analysis must be kept in mind while evaluating results.

Many workers have identified only large, or presumably adult, "typical" specimens in studies of populations of Foraminifera. This procedure automatically makes the population analyses qualitative, although they are frequently listed as quantitative data. The only valid quantitative approach is to count all specimens which occur in the fraction of the sample under study. In studying the complete population the range in morphological characters within highly variable species makes identification of some specimens difficult. Moreover, many small, juvenile specimens cannot be identified with certainty. The latter problem can be alleviated somewhat by counting two fractions in each population, specimens having a diameter greater than 0.15 mm. and those smaller than 0.15 mm. but greater than 0.062 mm., and these may be separated by standard sieves. This gives a separation of adults and juveniles except in very small species; the frequencies in each fraction may be determined separately since the identification problems are somewhat different.

The total population for each sample and the percentage frequency of each species or other group are calculated. From an ecological point of view the benthonic and planktonic species must be considered as two separate populations; both the planktonic and benthonic assemblages in each sample are listed as comprising 100%.

Listing frequencies as fractions of per cents does not seem to be valid on the basis of the accuracy of the data discussed above, and frequencies are given to the nearest whole per cent. Frequencies of less than 1% often are listed as 1%, although occasionally these may be given in tenths of one per cent. Similarly, the total populations are given to the nearest whole number, usually in hundreds, except for small populations where the total population is counted.

There has been discussion in the literature concerning the measurement of the size of samples from which quantitative studies of Foraminifera are made. Walton (1955, p. 992) has given an interesting discussion of this problem, as follows:

"There appear to be two schools of thought concerning the problem of equal samples, one favoring population counts from samples of equal dry weight (Schott, 1935; Said, 1950) and the other from equal volume samples (Phleger, 1952b; Höglund, 1947; Parker, 1948; Butcher, 1951).

"Most ecological work on modern Foraminifera is aimed at determining their distributions and relative abundances in modern sediments, which is a function of numerous and complex ecologic factors. Assuming that Foraminifera live at or just below the sediment-water interface, the natural measure of populations for comparative purposes should be the number of specimens per unit area. Assuming further that each environment is supporting the maximum number of Foraminifera possible, referring the population to any base other than available living space appears to be artificial. It is considered that wet volume of sediment is the only natural base to which living populations can be referred for comparative purposes.

"It can be shown that the reference of benthonic foraminiferal populations to constant dry-weight samples may be misleading. The total living population at station B-111 in 6 fathoms is 75 specimens, the wet volume is approximately 10 cc., and the total dry weight is 13.8 grams. Station B-45 in 141 fathoms has a total living population of 102 specimens from a volume of 10 cc. of wet sediment, but the total dry weight of this sample is 4.28 grams. These living populations are comparable when based on an equal volume sample, but based on equal dry weights the living population at station B-45 would be approximately 4 times as great as the population from sample B-111. This would suggest that at depths of 141 fathoms, the environment can support 4 times as many living Foraminifera as the shallow-water environment. Such conclusions are not valid in the light of what is known about the distribution of living Foraminifera.

"If the sediment samples have been washed or otherwise concentrated before the dry weight is measured, the data are even less reliable. The weights just listed for samples B-45 and B-111 include the entire sediment sample. If only the coarser fractions of the samples are considered (sizes greater than $3\frac{1}{2}$ ϕ or 0.088 mm.), which contains most of the Foraminifera, the dry weight of sample B-45 is 1.03 grams, and that of B-111 is 7.035 grams. It is obvious that population counts from dry, previously washed, or concentrated samples are not comparable and should not be considered representative."

Wet-volume samples from the modern ocean are not directly comparable to samples of ancient marine rocks. This may be resolved by measuring dry weight, in addition to volume. It is desirable to obtain sufficient data so that wet volume can be converted into dry weight of different types of sediment. Population data per unit area are available in many studies and these may be useful in application to ancient sediment.

Preparation of Samples. In the Marine Foraminifera Laboratory we have applied the usual well-known methods of preparation. In identifying living specimens, however, it is desirable to keep the sample in a water and formalin solution so that the cell material is retained intact. Several delicate species of Foraminifera, usually rare or absent in dried samples, have been discovered in abundance in samples which have never been dried. These are generally thin-walled, arenaceous forms and some chitinous types, and most of these specimens appear to disintegrate after drying.

In addition, some of the larger arenaceous species appear to be injured and occasionally destroyed by the drying process. We have observed that *Ammotium*, for example, may be abundant in certain samples studied in solution and usually will be much less abundant in the same samples after they are dried. Several dry samples have been examined which had previously been treated with a dispersing agent; although many of these samples were from environments which normally contain abundant arenaceous Foraminifera, no specimens of such forms have been observed. Certain marine sediments should be stored and examined in a wet condition; such a procedure may improve the ecologic interpretation of faunas in which arenaceous specimens occur.

CHAPTER II

Depth Distribution of Benthonic Foraminifera

General Statement

Depth is one of the obvious variables in the ocean. It has long been known that there is a depth zonation of larger marine benthonic organisms, but useful and reliable data are scant, with some exceptions (see R. H. Parker, 1960). Information on depth distribution of benthonic Foraminifera appears to be more reliable and abundant than for any other group of organisms. There is, however, a certain amount of apparent confusion about these depth distributions inasmuch as each worker has defined somewhat different depth zones. This may be due to actual environmental differences with depth in different areas or to different interpretations of similar data, or both. Most of the recent studies in this field are summarized in this chapter, generalizations on known depth distributions are attempted, the possible ecologic factors which cause such distributions are discussed, and uses of information on depth distributions are outlined.

A considerable amount of useful, quantitative data has been

accumulated on the depth distribution of benthonic species of Foraminifera. These data are from the northern Gulf of Mexico (Phleger, 1951b, Bandy, 1954 and 1956, Lowman, 1949, F. L. Parker, 1954), the Gulf of Maine (Phleger, 1952b), the western North Atlantic continental shelf (F. L. Parker, 1948, Phleger and Walton, 1950) and slope (Phleger, 1942), the eastern Atlantic (Colom, 1950 and 1952), off the coast of California (Natland, 1933, Butcher, 1951, Crouch, 1952, Bandy, 1953, Uchio, 1960), off Sweden (Höglund, 1947), in the Red Sea (Said, 1950), off the coast of Baja California (Walton, 1955), the coast of Asia (Polski, 1959), the Okhotsk Sea (Saidova, 1957a), and other Russian seas (Stschedrina, 1958a), and in the Mediterranean (F. L. Parker, 1958). The depth ranges shown in these different areas appear to fit into a general pattern.

In the northwestern Gulf of Mexico (Phleger, 1951b) the most marked faunal boundary is at a depth of approximately 100 m., and other deeper boundaries are reported at 200 m., 600 m., 1000 m., and 2000 m. water depth. None of these boundaries is sharp but each has a variability in depth of 10-20%. Lowman (1949) recognized a striking faunal change at a depth of approximately 300 ft. (91 m.) in the northern Gulf of Mexico.

Bandy (1954) has studied several samples from three inshore traverses west of the Mississippi Delta in the northern Gulf of Mexico, from depths of 27 ft. (8 m.) to 130 ft. (40 m.). He has recognized faunal depth boundaries at 55 ft. (17 m.) and 75 ft. (23 m.). Recently boundaries between foraminiferal assemblages have been reported from the continental shelf off southwest Texas (Phleger, 1956) at 20-30 m., 50-70 m., and 100 m., and there are suggestions of less distinct boundaries at intermediate depths. Depth zonation of these faunas is based on actual living populations.

Bandy (1956) recognizes boundaries between different faunas on the continental shelf in the northeastern Gulf of Mexico at 12 m., 32 m., 55 m., 76 m., and 122 m. F. L. Parker

(1954) has published foraminiferal distributions in the northeastern Gulf of Mexico between the Mississippi Delta and Florida. She recognizes the following boundaries: 80-100 m., 130-150 m., 180-220 m., 350-600 m., and 900-1000 m.

F. L. Parker (1948) in her study of Foraminifera faunas from the continental shelf between the coast of Maryland and Cape Ann also reports a distinct depth zonation of benthonic faunas. Her faunal boundaries are given as 90 m. and 300 m. Boundaries are suggested on the middle and lower part of the Atlantic continental slope at 600 m., 1000 m., and 1600 m. (Phleger, 1942). A marked benthonic faunal boundary occurs in the southern Gulf of Maine, off Portsmouth, New Hampshire, at approximately 60-75 m. (Phleger, 1952b). A boundary occurs at approximately 20 m. water depth in Cape Cod Bay (Phleger and Walton, 1950).

Colom (1952) has suggested three principal depth zones off the coast of Galicia, with boundaries at approximately 15 m. and 200 m. This author also shows a possible additional boundary at 400-500 m. Examination of Colom's range chart suggests the possibility that there may be other depth zones in this area, but study of additional material would be desirable to substantiate this.

Said (1950) reports faunal depth boundaries at 70 m., 300 m. and 500 m. in the Red Sea, based on thirty-seven samples.

Höglund (1947, pp. 294-295, figs. 311, 312) has summarized the bathymetric ranges of benthonic Foraminifera in the Skagerak and the Gullmar Fjord, off the west coast of Sweden, and has graphically illustrated twenty-four different types of bathymetric distributions. The generalized frequency curves shown on his diagrams strongly suggest faunal boundaries at approximately 200 m., 300 m. and 500 m. in the Skagerak, an open-ocean body of water. In the Gullmar Fjord there appear to be at least three faunal boundaries above 120 m., possibly at 20 m., 40 m. and 80 m.

Kruit (1955) has delineated a faunal boundary in the western Mediterranean off the Rhone Delta at 12-30 m. F. L.

Parker (1958) recognizes foraminiferal depth boundaries in the eastern Mediterranean at 143-205 m., 500-700 m. and 1000-1300 m.

A fauna has been recognized in Baffin Bay off Greenland (Phleger, 1952a) which appears to be generally indicative of water depths greater than 250 m.; this possible zonation is based on very few samples, and there is faunal contamination in the area due to transport of sediment by sea ice. Carsola (1952, pp. 170-177) has tentatively suggested benthonic Foraminifera depth boundaries at 65 m. and 450 m. from the Arctic Ocean off Alaska and northwestern Canada, and Loeblich and Tappan (1953) suggest a boundary at about 50 m.

Natland's (1933) study of depth distribution of Foraminifera off the coast of southern California shows depth boundaries at 125 ft. (38 m.), 900 ft. (274 m.) and 6500 ft. (1980 m.). Bandy (1953) has recognized depth boundaries at 150 ft. (50 m.), 600 ft. (200 m.), 6000 ft. (2000 m.) and 8000 ft. (2500 m.). Butcher (1951), in a study of faunas off San Diego, has an irregular major faunal boundary at 290-440 m., and others at 130-260 m. and 730-810 m.; none of Butcher's samples was from depths less than 100 m. Crouch (1952) lists Foraminifera depth boundaries at 900 ft. (274 m.), 2000 ft. (610 m.), 4000 ft. (1219 m.) and 7500 ft. (2286 m.).

Walton (1955) has determined the following depth boundaries in the area of Todos Santos Bay, Baja California: 30 fm. (55 m.), 50 fm. (91 m.), 100 fm. (183 m.) and 350 fm. (640 m.). His collections extended only to a depth of 600 fm. (1097 m.).

Uchio (1960) in his study of samples along several traverses off San Diego, California, has established faunal boundaries at 13 fm. (24 m.), 45 fm. (82 m.), 100 fm. (183 m.), 250 fm. (457 m.), 350 fm. (640 m.) and 450 fm. (823 m.).

Stschedrina (1958a) has summarized studies of depth distribution of species in the Arctic and northwest Pacific adjacent to the U.S.S.R. Her sublittoral fauna is at less than 50 m. and is composed of "cold water forms." A eulittoral fauna is at 50-200 m., upper bathyal 200-750 m., lower bathyal 750-2000 m., abyssal 2000-7000 m. and ultra-abyssal deeper than 7000 m. Polski (1959) studied faunas from the north Asiatic coast, principally the East China Sea. There is an inner shelf fauna less than 25 fm. (46 m.), central shelf 25-43 fm. (46-79 m.), outer shelf 43-66 fm. (79-121 m.), upper bathyal 66-333 fm. (121-609 m.) and middle bathyal fauna deeper than 333 fm. (609 m.).

GENERALIZED DEPTH RANGES

Figure 13 summarizes much of the published information on depth assemblages of Foraminifera. The reliability of this information varies, since in some cases the depth distributions are based on inadequate or somewhat questionable data. Some of the depth boundaries listed in Figure 13 are the present writer's interpretation of data given. Stschedrina (1957) also has summarized several authors' depth data, as has F. L. Parker (1958).

Certain generalizations may be made from the list in Figure 13. There seem to be rather widespread faunal boundaries at approximately 20 m., 50 m., 100 m., 200-300 m., 400-500 m., 1000 m. and 2000 m. The boundary at 2000 m. is not so well-known as the others because few faunas have been studied from such relatively great depths. Stschedrina (1958a) also suggests an ultra-abyssal boundary at 7000 m. The most marked faunal boundary seems to be at approximately 100 m. in the northern Gulf of Mexico and off the North American Atlantic coast south of Cape Cod. In the southern Gulf of Maine it is at about 70 m. Off northern Baja California, Walton (1955) shows marked boundaries at about 50 m. and 100 m., and these approximately correspond to those reported by Natland (1933) and Uchio (1960) from southern California. The

GULF OF MEXICO						W.N. ATLANTIC		E.N. ATLANTIC		MEDIT.	RED SEA	SWEDEN	SOUTHERN CALIFORNIA					CENTRAL AMERICA	ASIA	ARCTIC	
LOWMAN 1949	BANDY 1954	PHLEGER 1956	PHLEGER 1951 b	BANDY 1956	F.L.PARKER 1954	F.L.PARKER 1948	PHLEGER 1942	COLOM 1952	COLOM 1950	F.L.PARKER 1958	SAID 1950	HOGLAND 1947	NATLAND 1933	CROUCH 1952	BANDY 1953	UCHIO 1960	WALTON 1955	BANDY & ARNAL 1957	POLSKI STSCHEDRINA 1959 1958a	CARSOLA 1952	
"NERITIC"	1												I		MIDDLE NERITIC	1		INNER SHELF	INNER SHELF SUB-LITTORAL	I	
	2		1	1		1				1											
	3			2									II			2			CENTRAL SHELF		
				3		2									LOWER NERITIC			OUTER SHELF	OUTER SHELF EU-LITTORAL		
				4						2	2		III			3					
			2	5	2																
					3	3														II	
								?		3	3							UPPER BATHYAL	UPPER BATHYAL		
"BATHYAL"			3		4	4							IV	T 4	BATHYAL	4					
										4	4										
			4		5					5				T 5		5		MIDDLE BATHYAL	MIDDLE BATHYAL	III	
			5											T 6		6		LOWER BATHYAL	LOWER BATHYAL		
			6		6								V	T 7	UPPER ABYSSAL LOWER ABYSSAL						
																			ABYSSAL		

DEPTH IN METERS: 0, 50, 100, 200, 300, 400, 500, 1000, 3000

depth of this marked boundary in these areas appears to coincide with the depth of the seasonal layer of the ocean water at the places where it is observed. This and other possible explanations for depth zonation are discussed in a later section of this chapter.

Depth Ranges of Species

It is not possible at the present time to generalize on the world-wide depth distributions of many actual species or other taxonomic groups of Foraminifera because there has been insufficient study of faunas from this particular point of view. Also, F. L. Parker (personal communication) has pointed out that probably there are few benthonic species which are worldwide in distribution except those in the deep sea. Many of the samples which have been studied are of questionable reliability. Identification of species varies with authors and it is difficult to compare faunal lists if no illustrations of species are included or no references are given to illustrations. In some instances forms reported as characteristic of deep water actually have been displaced from shallower water, both by "turbidity currents" and ice rafting. Another type of "natural contamination" may come from erosion of submarine fossil exposures, as in the continental borderland off southern California, or by residual assemblages which have not been covered subsequently with sediment. The only reliable depth distributions are based on living assemblages. These possibilities must be considered in evaluating any ecological study such as depth assemblages and especially in comparing distributions

---

Figure 13, Opposite. Summary of depth assemblages of benthonic Foraminifera reported from various areas of the world.

in different areas. Finally, few studies have been made on a quantitative basis; qualitative results are difficult to compare with quantitative data.

Three types of species distributions may be used in the delineation of depth biofacies of Foraminifera, as follows:

1. Discrete depth ranges, in which species have a definite shallow and/or deep limit of occurrence. These are best developed on the continental shelf, e.g., in the seasonal layer in mid-latitudes, where many species do not extend deeper than 70-100 m. and their shallow limit is the shoreline. Shallow range limits are considered to be more reliable than deep limits.

2. Overlap ranges of species, in which a depth facies can be recognized by the occurrence of two or more species together only within a specific depth range.

3. Frequencies of species often are useful. Some forms have a markedly higher frequency within a specific depth range even though they may be present at virtually all depths.

Some species appear to occur in many environments at about the same frequency in some areas and thus have no value for delineation of different environments. It is necessary that a species be sufficiently common in most of the samples within its environmental limits to be useful. Species which occur rarely in a few samples are of limited value since their presence or absence may be misleading. The characteristic of a generally rare species is that it may occur in occasional samples in moderate abundance but be absent at most stations, and it may be difficult to ascertain its real distribution. It appears desirable generally to disregard such forms in the final delineation of a modern association. It is always possible, however, that occasional rare forms may be of considerable value in interpreting ancient rocks.

In general, species rather than larger groups have been used in modern assemblage diagnoses, although there are some exceptions. The desirability of using species for such a purpose may be illustrated by the depth distributions of various species of *Bolivina* in the northwestern Gulf of Mexico (see Figures 14-17). *B. pulchella primitiva* Cushman and *B.* sp. are confined to depths less than 100 m.; *B. striatula spinata* Cushman is most common at less than 100 m. but occurs to a depth of 200 m.; *B. goësii* Cushman is recorded only from 100-200 m.; *B. hastata* Phleger and Parker occurs between 100 and 400 m.; *B. ordinaria* Phleger and Parker is most abundant between 200 m. and 1000 m.; *B. barbata* Phleger and Parker occurs from 50 m. to 1000 m. but is most abundant at 100-200 m.; and *B. lowmani* Phleger and Parker occurs in significant frequencies at all depths from 20 m. to 3500 m. These examples are sufficient to demonstrate that the consideration of the genus *Bolivina* without species subdivision could give an incomplete depth distribution analysis. The use of species adds difficulty to detailed faunal comparisons between regions since there may be different speciation. Further comparative studies may reveal species or subspecies similarities which are useful in several areas.

WORLD-WIDE DEPTH DISTRIBUTIONS OF SPECIES

Depth ranges of all significant species from the areas where apparently reliable studies have been made have been tabulated by the writer. The only generalizations on widespread distributions of species which appear to be possible from these data are as follows:

1. The following appear to be characteristic of water depths less than approximately 70-100 m.:

> *Buliminella elegantissima* (d'Orbigny)
> some species of *Rosalina*
> some species of *Eggerella*
> *Elphidium* spp.

(Text continued on page 52.)

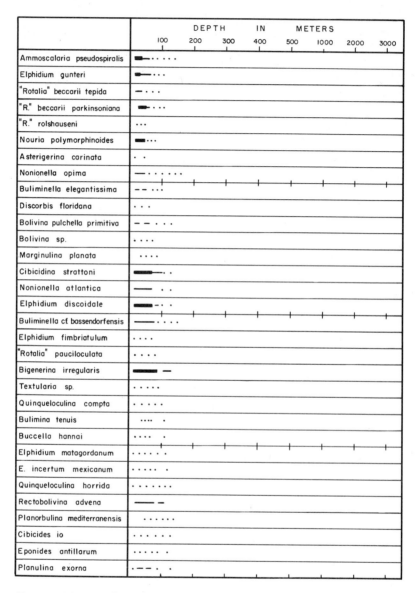

Figure 14. Depth ranges of the more abundant species of Foraminifera in the northwestern Gulf of Mexico. Relative abundance shown by heaviness of line.

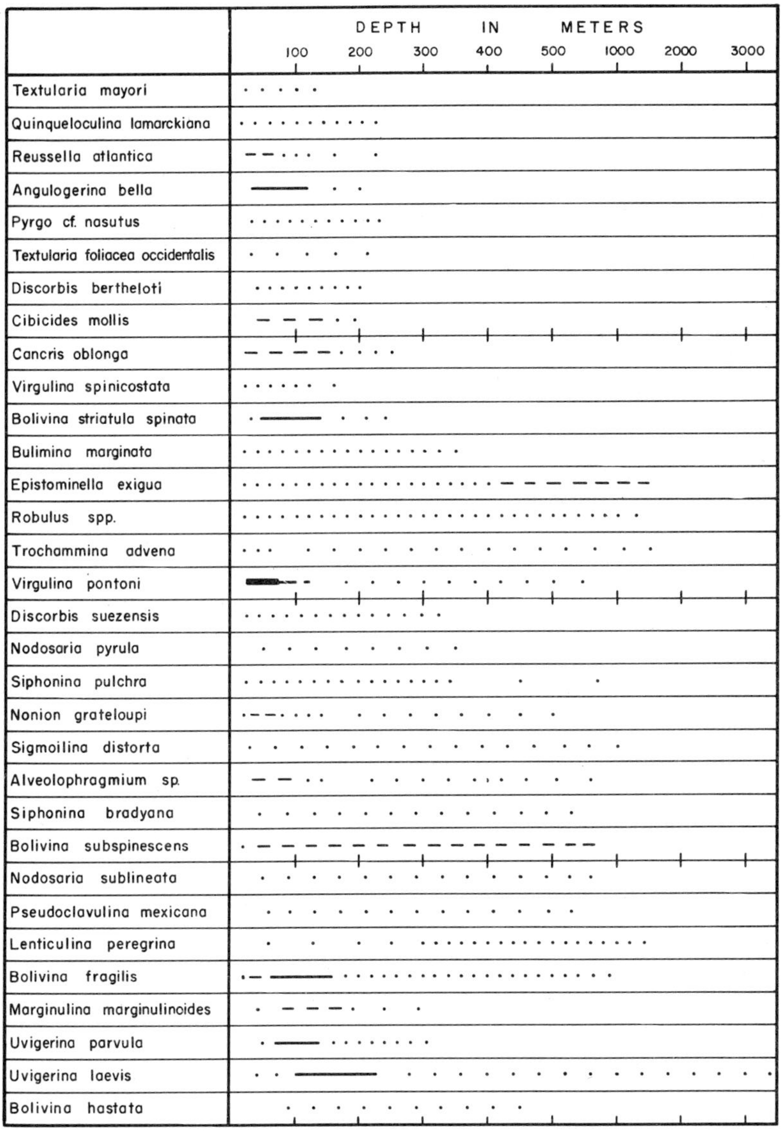

	DEPTH IN METERS 100 200 300 400 500 1000 2000 3000
Textularia mayori	
Quinqueloculina lamarckiana	
Reussella atlantica	
Angulogerina bella	
Pyrgo cf. nasutus	
Textularia foliacea occidentalis	
Discorbis bertheloti	
Cibicides mollis	
Cancris oblonga	
Virgulina spinicostata	
Bolivina striatula spinata	
Bulimina marginata	
Epistominella exigua	
Robulus spp.	
Trochammina advena	
Virgulina pontoni	
Discorbis suezensis	
Nodosaria pyrula	
Siphonina pulchra	
Nonion grateloupi	
Sigmoilina distorta	
Alveolophragmium sp.	
Siphonina bradyana	
Bolivina subspinescens	
Nodosaria sublineata	
Pseudoclavulina mexicana	
Lenticulina peregrina	
Bolivina fragilis	
Marginulina marginulinoides	
Uvigerina parvula	
Uvigerina laevis	
Bolivina hastata	

Figure 15. Depth ranges of the more abundant species of Foraminifera in the northwestern Gulf of Mexico. Relative abundance shown by heaviness of line.

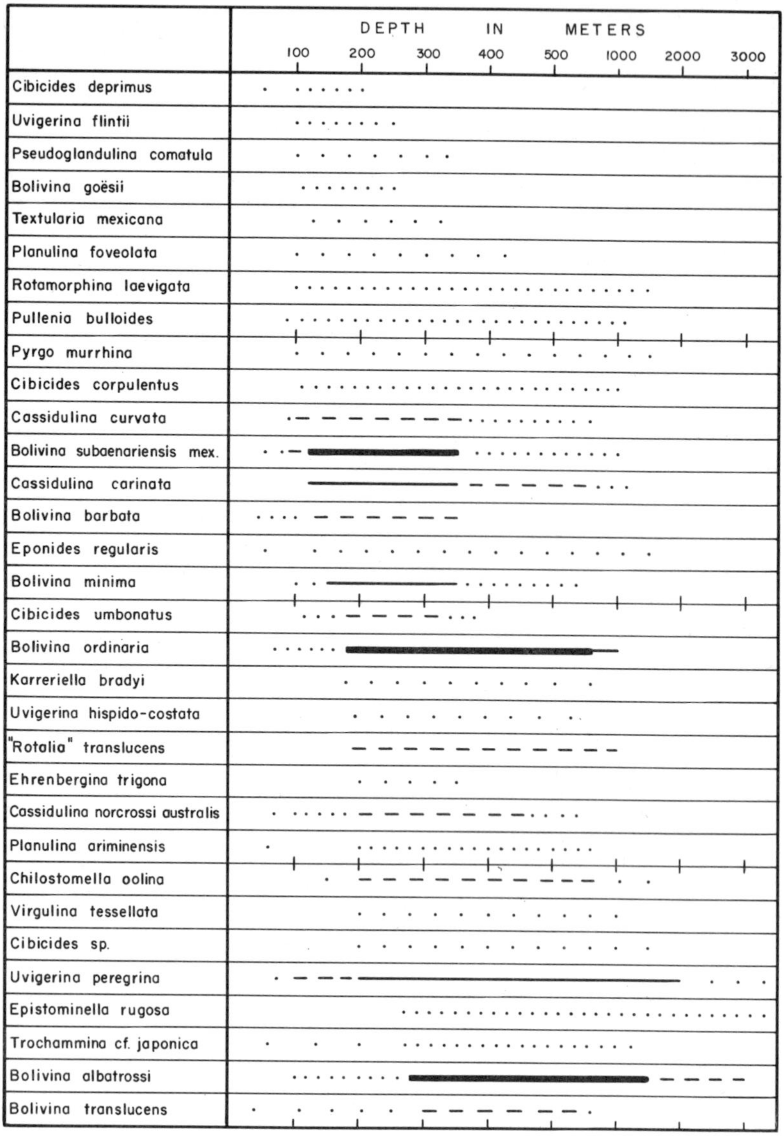

50 Ecology and Distribution of Recent Foraminifera

	DEPTH IN METERS							
	100	200	300	400	500	1000	2000	3000
Cibicides deprimus								
Uvigerina flintii								
Pseudoglandulina comatula								
Bolivina goësii								
Textularia mexicana								
Planulina foveolata								
Rotamorphina laevigata								
Pullenia bulloides								
Pyrgo murrhina								
Cibicides corpulentus								
Cassidulina curvata								
Bolivina subaenariensis mex.								
Cassidulina carinata								
Bolivina barbata								
Eponides regularis								
Bolivina minima								
Cibicides umbonatus								
Bolivina ordinaria								
Karreriella bradyi								
Uvigerina hispido-costata								
"Rotalia" translucens								
Ehrenbergina trigona								
Cassidulina norcrossi australis								
Planulina ariminensis								
Chilostomella oolina								
Virgulina tessellata								
Cibicides sp.								
Uvigerina peregrina								
Epistominella rugosa								
Trochammina cf. japonica								
Bolivina albatrossi								
Bolivina translucens								

Figure 16. Depth ranges of the more abundant species of Foraminifera in the northwestern Gulf of Mexico. Relative abundance shown by heaviness of line.

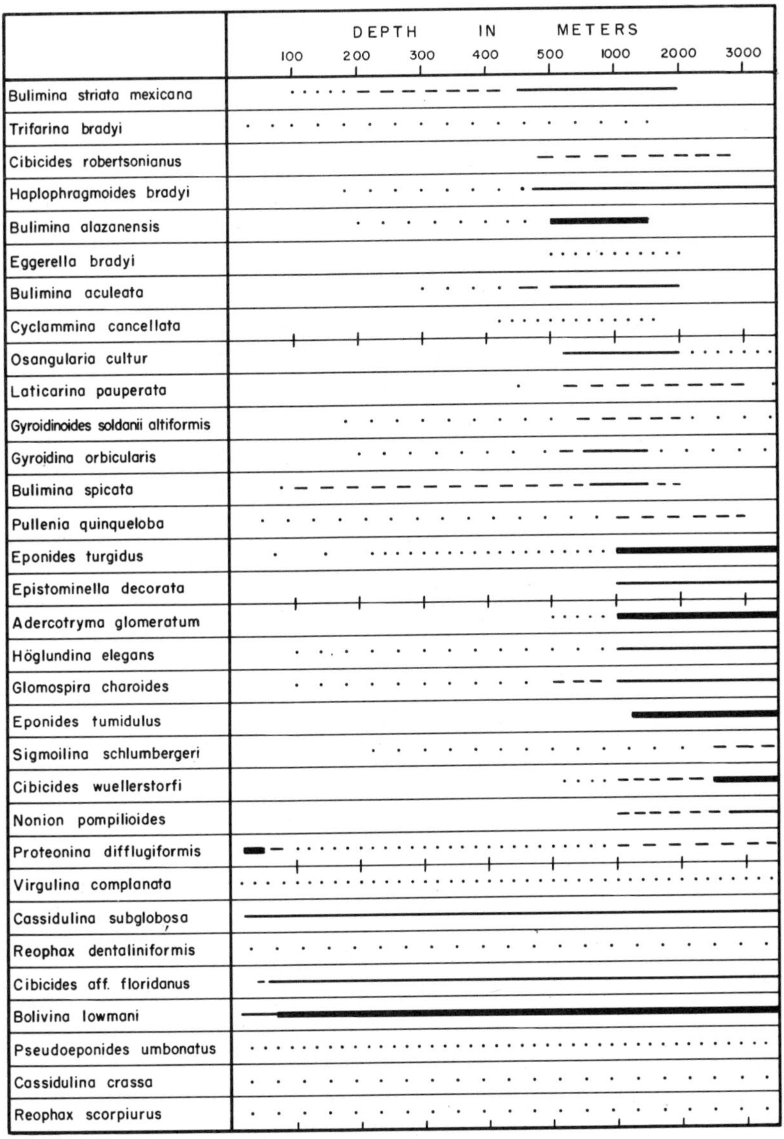

CHAPTER II 51

Species	DEPTH IN METERS 100 200 300 400 500 1000 2000 3000
Bulimina striata mexicana	
Trifarina bradyi	
Cibicides robertsonianus	
Haplophragmoides bradyi	
Bulimina alazanensis	
Eggerella bradyi	
Bulimina aculeata	
Cyclammina cancellata	
Osangularia cultur	
Laticarina pauperata	
Gyroidinoides soldanii altiformis	
Gyroidina orbicularis	
Bulimina spicata	
Pullenia quinqueloba	
Eponides turgidus	
Epistominella decorata	
Adercotryma glomeratum	
Höglundina elegans	
Glomospira charoides	
Eponides tumidulus	
Sigmoilina schlumbergeri	
Cibicides wuellerstorfi	
Nonion pompilioides	
Proteonina difflugiformis	
Virgulina complanata	
Cassidulina subglobosa	
Reophax dentaliniformis	
Cibicides aff. floridanus	
Bolivina lowmani	
Pseudoeponides umbonatus	
Cassidulina crassa	
Reophax scorpiurus	

Figure 17. Depth ranges of the more abundant species of Foraminifera in the northwestern Gulf of Mexico. Relative abundance shown by heaviness of line.

Miliolidae in abundance
Quinqueloculina seminulum (Linné) variants
Streblus beccarii (Linné) variants

2. The following seem to be characteristic of depths greater than approximately 100 m.:

Höglundina elegans (d'Orbigny)
Uvigerina peregrina Cushman

Akers (1954) recently has pointed out that *Cyclammina cancellata* Brady is restricted to depths in the modern seas greater than approximately 200 m. This and closely related species are abundant in some Tertiary rocks. This form is only locally abundant in modern marine sediments and is of so little quantitative importance that it has been largely ignored. There may be other relatively rare, modern forms which have equivalent ecologic value in older rocks but which also have not been emphasized sufficiently in modern ecologic studies.

LOCAL DEPTH DISTRIBUTIONS

The depth ranges of significant species in many of the following areas are summarized in Figures 14 to 36: the southwestern Gulf of Maine (Phleger, 1952b), the western Atlantic continental shelf (F. L. Parker, 1948), the northwestern Gulf of Mexico (Phleger, 1951b, 1956), the northeastern Gulf of Mexico (F. L. Parker, 1954) and off Baja California, Mexico (Walton, 1955). Illustrations of some of the species associations in depth biofacies for the northwestern Gulf of Mexico and the southern Gulf of Maine are shown on Plates 1-6.

Recently depth occurrence data on benthonic Foraminifera from the northwestern Gulf of Mexico (Phleger, 1951b) have been re-analyzed. Four groups of relative frequencies are indicated by the heaviness of the range lines in Figures 14-17. These arbitrary frequency groups were established by considering for each species: (1) the percentage frequencies, (2) the percentage of the stations at which it occurred, (3) the number of traverses in which it occurred, and (4) occurrences of living specimens. The range lines shown on Figures 14-17 were determined by inspection of these data for each species plotted on sheets which record the variables indicated. No attempt was made to reduce the data to a specific formula because such a procedure would imply a greater accuracy in collection and study of the samples than seems to be warranted. It is believed that the range lines showing relative frequencies are a good representation of depth distributions in the samples from this area.

There are some rather interesting generalizations which may be made from these depth distributions. In the first place, these new analyses show that the faunal depth boundaries already published are essentially correct. But the data show, in addition, that many more "sub-boundaries" exist than were realized at the time of the original studies. Each species has its own characteristic distribution which seems to be different from those of most or all other species. This suggests the possibility that detailed depth zonation can be established on the basis of such data. It is apparent that there are several faunal depth boundaries on the continental shelf.

The presence of additional depth zones on the northwest Gulf of Mexico continental shelf is confirmed by study of closely-spaced samples in some shelf traverses off the San Antonio Bay area in Texas (Phleger, 1956). The ranges of living and total populations are shown on Figures 18-21, with relative abundance indicated by heaviness of the lines. Faunal boundaries at 20-30 m., 50-70 m., and 90-100 m. are marked. The total (live plus dead) populations extend deeper than the living population in many species; the shallow limits of the more abundant species are essentially similar. Boundaries are not sharp but are gradational. These depth distributions are considered to have a high degree of reliability since they are largely based on living assemblages.

(Text continued on page 80.)

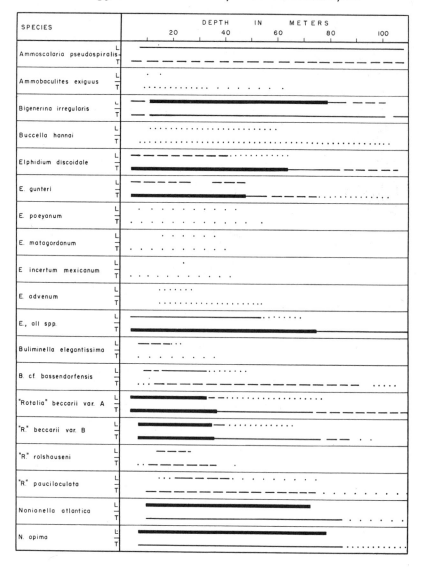

Figure 18. Generalized depth ranges of benthonic Foraminifera in continental shelf traverses off the central Texas coast. L = living population, T = total population. Heaviness of line indicates relative abundance. After Phleger (1956).

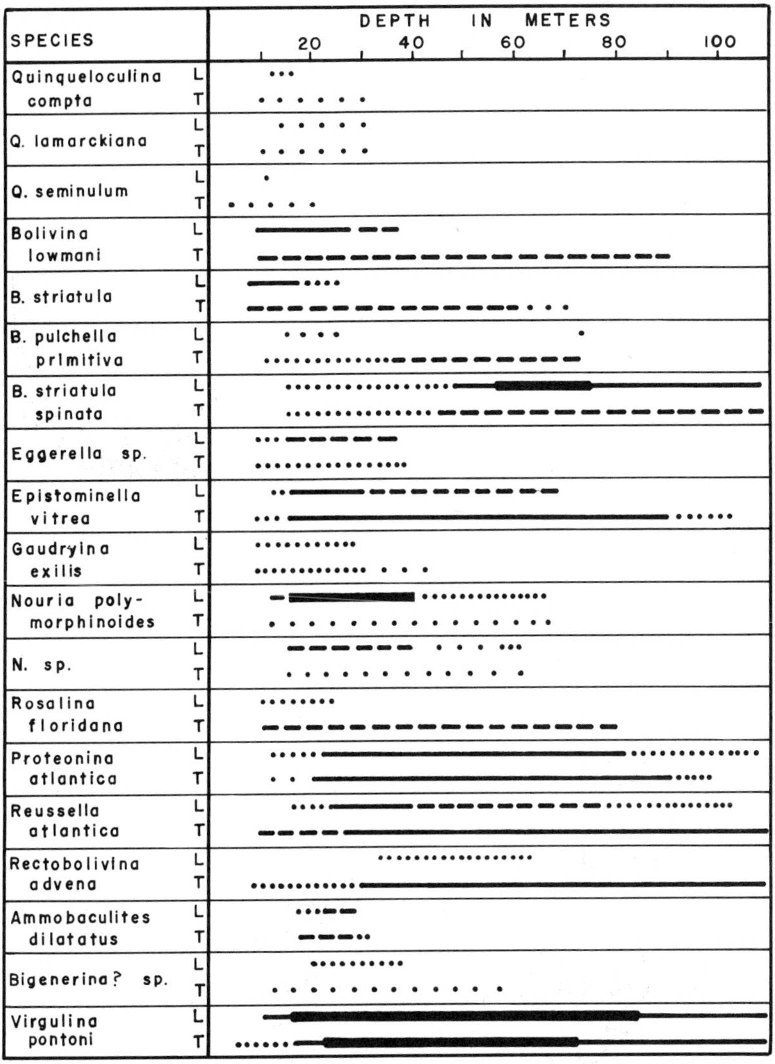

SPECIES		DEPTH IN METERS 20 40 60 80 100
Quinqueloculina compta	L T	
Q. lamarckiana	L T	
Q. seminulum	L T	
Bolivina lowmani	L T	
B. striatula	L T	
B. pulchella primitiva	L T	
B. striatula spinata	L T	
Eggerella sp.	L T	
Epistominella vitrea	L T	
Gaudryina exilis	L T	
Nouria poly- morphinoides	L T	
N. sp.	L T	
Rosalina floridana	L T	
Proteonina atlantica	L T	
Reussella atlantica	L T	
Rectobolivina advena	L T	
Ammobaculites dilatatus	L T	
Bigenerina? sp.	L T	
Virgulina pontoni	L T	

Figure 19. Generalized depth ranges of benthonic Foraminifera in continental shelf traverses off the central Texas coast. L = living population, T = total population. Heaviness of line indicates relative abundance. After Phleger (1956).

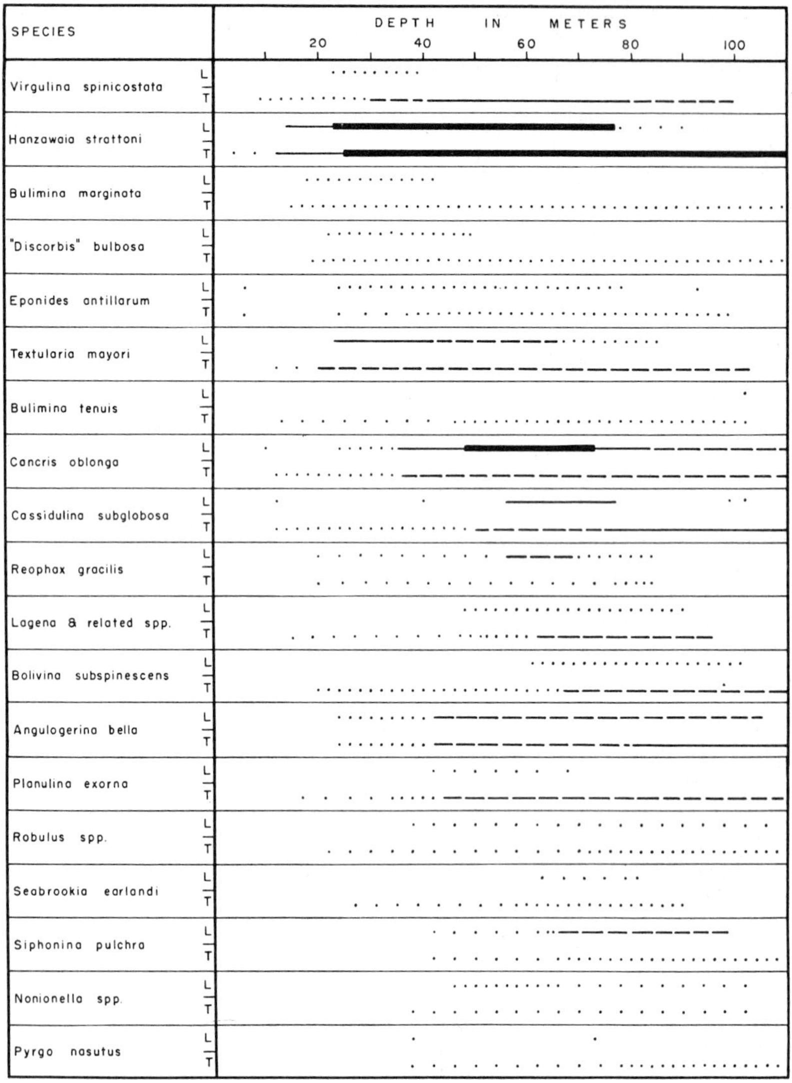

56 Ecology and Distribution of Recent Foraminifera

SPECIES		DEPTH IN METERS 20 40 60 80 100
Virgulina spinicostata	L / T	
Hanzawaia strattoni	L / T	
Bulimina marginata	L / T	
"Discorbis" bulbosa	L / T	
Eponides antillarum	L / T	
Textularia mayori	L / T	
Bulimina tenuis	L / T	
Cancris oblonga	L / T	
Cassidulina subglobosa	L / T	
Reophax gracilis	L / T	
Lagena & related spp.	L / T	
Bolivina subspinescens	L / T	
Angulogerina bella	L / T	
Planulina exorna	L / T	
Robulus spp.	L / T	
Seabrookia earlandi	L / T	
Siphonina pulchra	L / T	
Nonionella spp.	L / T	
Pyrgo nasutus	L / T	

Figure 20. Generalized depth ranges of benthonic Foraminifera in continental shelf traverses off the central Texas coast. L = living population, T = total population. Heaviness of line indicates relative abundance. After Phleger (1956).

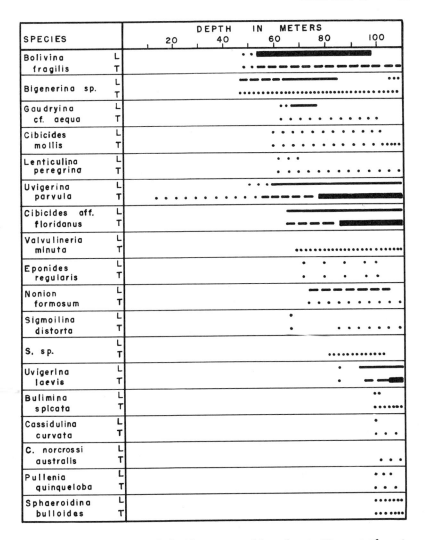

Figure 21. Generalized depth ranges of benthonic Foraminifera in continental shelf traverses off the central Texas coast. L = living population, T = total population. Heaviness of line indicates relative abundance. After Phleger (1956).

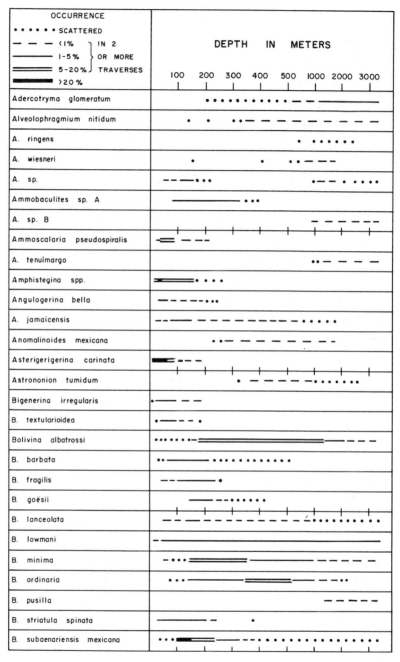

Figure 22. Depth ranges of the principal species of Foraminifera in the northeastern Gulf of Mexico. After F. L. Parker (1954).

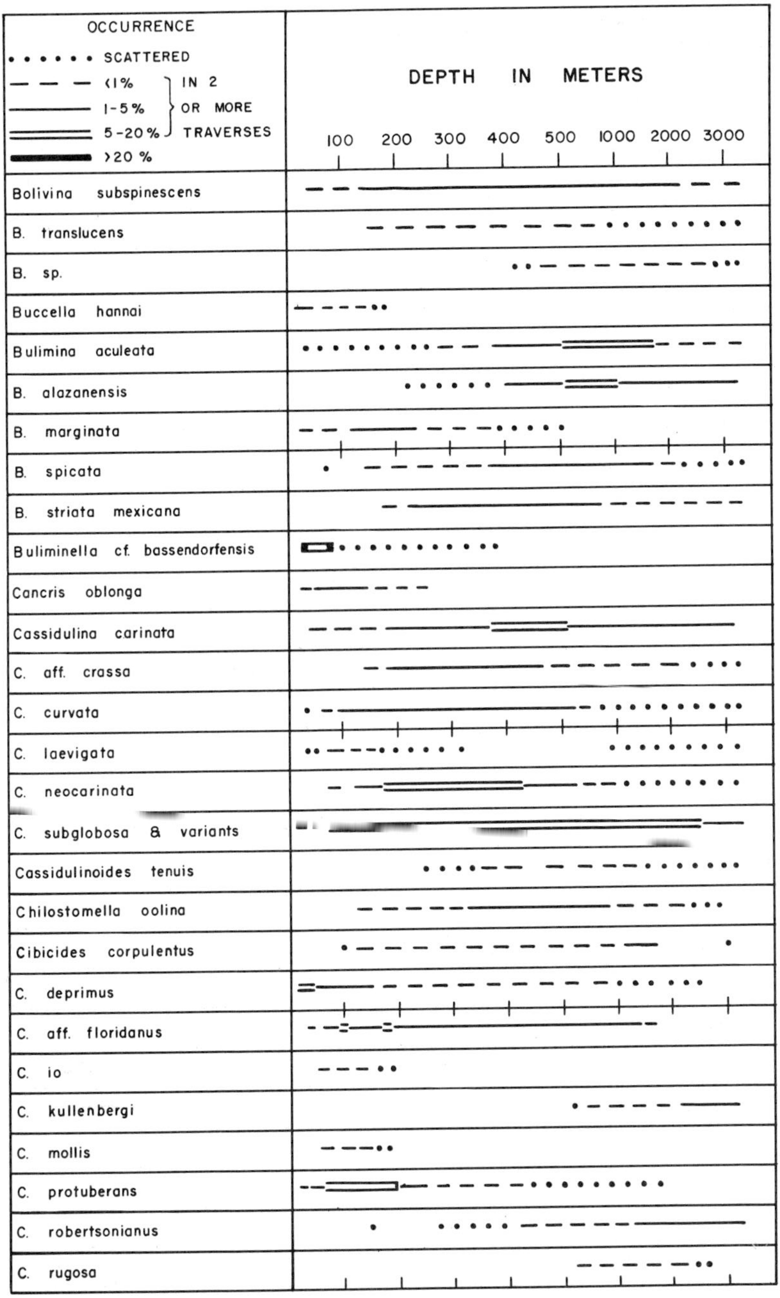

OCCURRENCE	DEPTH IN METERS
• • • • • SCATTERED — — — <1% ⎤ IN 2 ———— 1-5% ⎬ OR MORE ══════ 5-20% ⎦ TRAVERSES ▬▬▬▬ >20%	100 200 300 400 500 1000 2000 3000
Bolivina subspinescens	
B. translucens	
B. sp.	
Buccella hannai	
Bulimina aculeata	
B. alazanensis	
B. marginata	
B. spicata	
B. striata mexicana	
Buliminella cf. bassendorfensis	
Cancris oblonga	
Cassidulina carinata	
C. aff. crassa	
C. curvata	
C. laevigata	
C. neocarinata	
C. subglobosa & variants	
Cassidulinoides tenuis	
Chilostomella oolina	
Cibicides corpulentus	
C. deprimus	
C. aff. floridanus	
C. io	
C. kullenbergi	
C. mollis	
C. protuberans	
C. robertsonianus	
C. rugosa	

Figure 23. Depth ranges of the principal species of Foraminifera in the northeastern Gulf of Mexico. After F. L. Parker (1954).

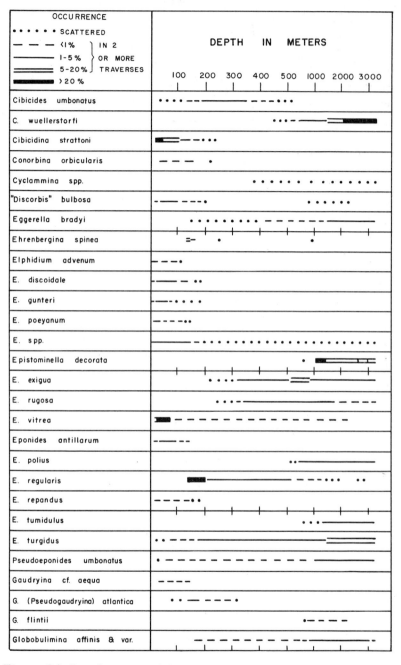

Figure 24. Depth ranges of the principal species of Foraminifera in the northeastern Gulf of Mexico. After F. L. Parker (1954).

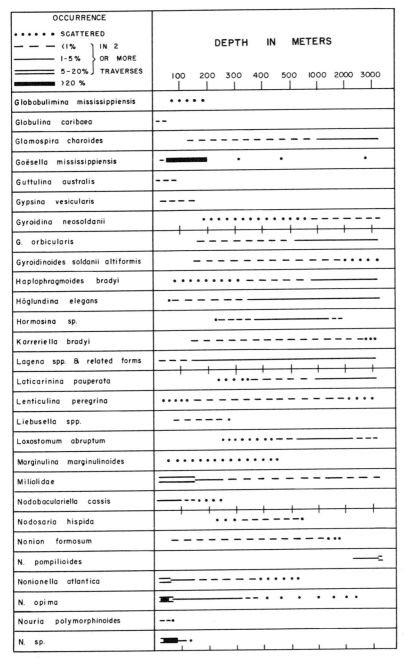

Figure 25. Depth ranges of the principal species of Foraminifera in the northeastern Gulf of Mexico. After F. L. Parker (1954).

Figure 26. Depth ranges of the principal species of Foraminifera in the northeastern Gulf of Mexico. After F. L. Parker (1954).

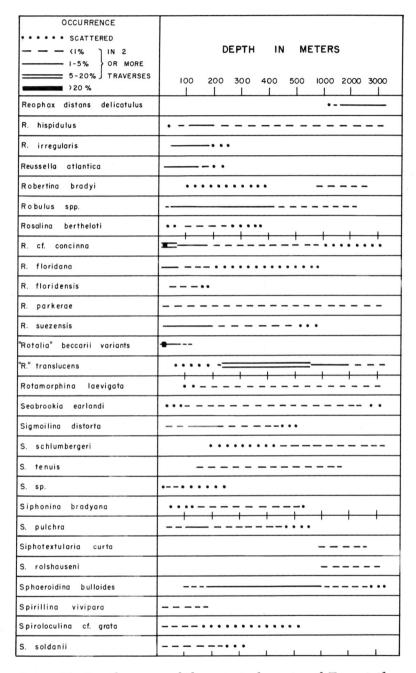

Figure 27. Depth ranges of the principal species of Foraminifera in the northeastern Gulf of Mexico. After F. L. Parker (1954).

Figure 28. Depth ranges of the principal species of Foraminifera in the northeastern Gulf of Mexico. After F. L. Parker (1954).

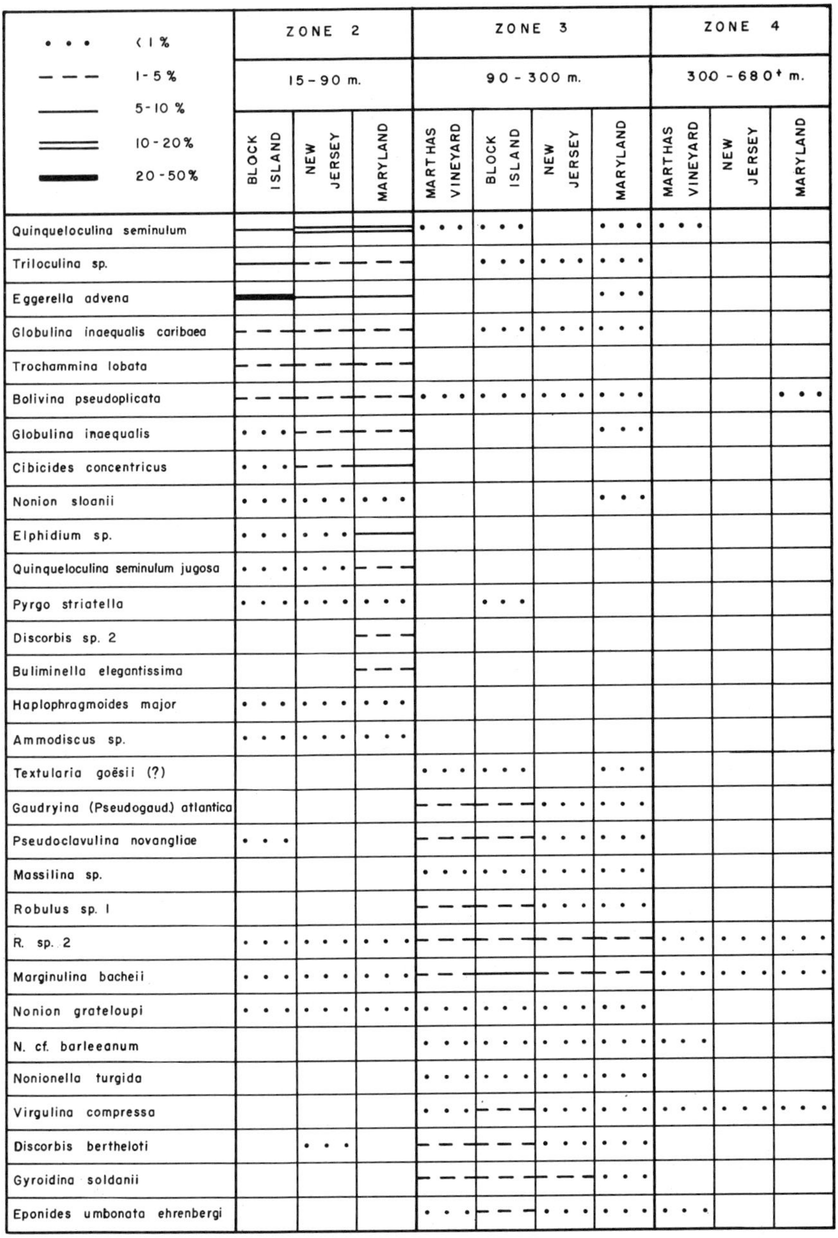

		Zone 2 15–90 m.			Zone 3 90–300 m.				Zone 4 300–680+ m.		
Legend		Block Island	New Jersey	Maryland	Marthas Vineyard	Block Island	New Jersey	Maryland	Marthas Vineyard	New Jersey	Maryland
...	<1%										
---	1–5%										
—	5–10%										
=	10–20%										
■	20–50%										
Quinqueloculina seminulum			=		...	...		...	...		
Triloculina sp.			---	---		...	...	...			
Eggerella advena		■						...			
Globulina inaequalis caribaea		---	---	---		...	...	...			
Trochammina lobata		---	---	---							
Bolivina pseudoplicata		---	---	---	...	...	---	...			...
Globulina inaequalis		...	---	---				...			
Cibicides concentricus		...	---								
Nonion sloanii		...	...	...				...			
Elphidium sp.		...	...								
Quinqueloculina seminulum jugosa		...	...	---							
Pyrgo striatella		...	...	...		...					
Discorbis sp. 2				---							
Buliminella elegantissima				---							
Haplophragmoides major		...	...	...							
Ammodiscus sp.		...	...	...							
Textularia goësii (?)					...	...		...			
Gaudryina (Pseudogaud.) atlantica					---	---	...	...			
Pseudoclavulina novangliae		...			---	---	...	...			
Massilina sp.					...	...	...	...			
Robulus sp. 1					---	---	...	...			
R. sp. 2		...	...	...	---	---	---	---	...	...	...
Marginulina bacheii		...	...	...	---			---	...	...	...
Nonion grateloupi		...	...	...	...	...	...	...			
N. cf. barleeanum					...	...	...	...	...		
Nonionella turgida					...	...	...	...			
Virgulina compressa					...	---	...	...	...	...	...
Discorbis bertheloti			...		---	---	...	...			
Gyroidina soldanii					---	---	---	...			
Eponides umbonata ehrenbergi					...	---	...	...	...		

Figure 29. Depth ranges of the principal species of Foraminifera on the Atlantic continental shelf from the Gulf of Maine to Maryland. After F. L. Parker (1948).

Legend		Zone 2 (15–90 m.)			Zone 3 (90–300 m.)				Zone 4 (300–680⁺ m.)		
• • •	< 1 %										
– – –	1-5 %										
——	5-10 %										
═══	10-20 %										
▬▬▬	20-50 %										

Species	Zone 2: Block Island	Zone 2: New Jersey	Zone 2: Maryland	Zone 3: Marthas Vineyard	Zone 3: Block Island	Zone 3: New Jersey	Zone 3: Maryland	Zone 4: Marthas Vineyard	Zone 4: New Jersey	Zone 4: Maryland
Epistomina elegans		• • •		• • •	– – –	– – –	• • •	• • •	• • •	
Cassidulinoides bradyi		• • •		• • •	• • •	• • •	• • •	• • •	• • •	• • •
Planulina foveolata		• • •		– – –	– – –	– – –	• • •			
Cancris sagra				• • •	• • •		• • •			
Eponides sp.				– – –	• • •	• • •	• • •			
Textularia pseudotrochus				• • •	• • •	• • •	• • •			
T. candeiana				• • •	• • •	• • •	• • •			
Karreriella affinis				• • •	• • •	• • •	• • •			
Pyrgo subsphaerica				• • •	• • •	• • •	• • •			
P. cf. denticulata		• • •	• • •	• • •	• • •		• • •			
Trochammina globigeriniformis				• • •	• • •	• • •	• • •			
Pseudopolymorphina atlantica		• • •	• • •	• • •	• • •	• • •	• • •			
Pulvinulinella cf. exigua			• • •	• • •		• • •	– – –	• • •	• • •	• • •
Chilostomella ovoidea				• • •	• • •		• • •			
Cassidulina laevigata			• • •	——	——	——	– – –	– – –	– – –	– – –
C. sp.			• • •	– – –	• • •	——	– – –		• • •	• • •
Listerella nodulosa							• • •	▬▬▬	– – –	– – –
Bulimina marginata var.							• • •	▬▬▬	═══	• • •
Gyroidina orbicularis				• • •			• • •	– – –	• • •	• • •
Valvulineria sp.							• • •	– – –	– – –	
Angulogerina angulosa var.								– – –	• • •	
Uvigerina peregrina								– – –	• • •	
Cassidulina norcrossi	– – –	• • •		– – –	– – –	• • •	– – –	═══	═══	——
Valvulina oonica								• • •	• • •	
Ammosphaeroidina sphaeroidiniform.								• • •	• • •	
Karreriella bradyi								• • •	• • •	
Pseudoglandulina occidentalis								• • •	• • •	
Nonion grateloupi var.							• • •	• • •	• • •	
Bolivina subspinescens								• • •	• • •	

Figure 30. Depth ranges of the principal species of Foraminifera on the Atlantic continental shelf from the Gulf of Maine to Maryland. After F. L. Parker (1948).

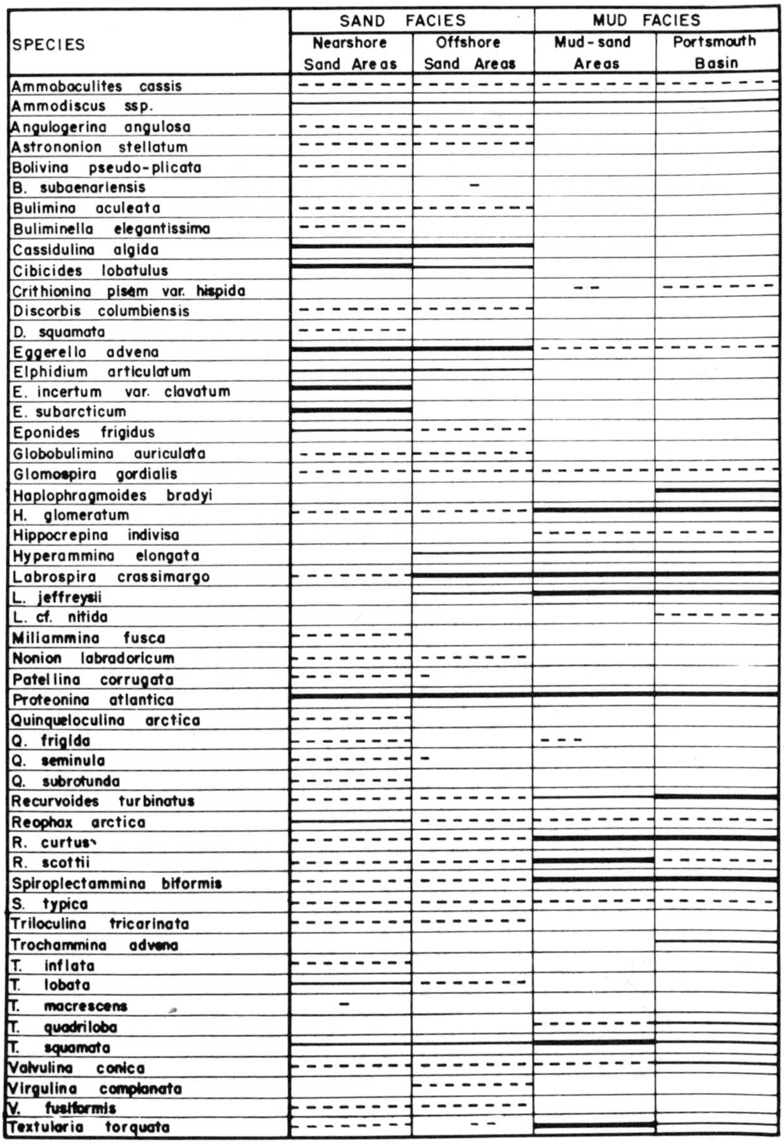

SPECIES	SAND FACIES		MUD FACIES	
	Nearshore Sand Areas	Offshore Sand Areas	Mud-sand Areas	Portsmouth Basin
Ammobaculites cassis	− − − − − −	− − − − − −	− − − − − −	− − − − − −
Ammodiscus ssp.				
Angulogerina angulosa	− − − − − −	− − − − − −		
Astrononion stellatum	− − − − − −	− − − − − −		
Bolivina pseudo-plicata	− − − − −			
B. subaenariensis		−		
Bulimina aculeata	− − − − − −	− − − − − −		
Buliminella elegantissima	− − − − − −			
Cassidulina algida				
Cibicides lobatulus	▬▬▬▬▬▬			
Crithionina pisum var. hispida			− −	− − − − − −
Discorbis columbiensis	− − − − − −			
D. squamata	− − − − − −			
Eggerella advena	▬▬▬▬▬▬		− − − − − −	− − − − − −
Elphidium articulatum	▬▬▬▬▬▬			
E. incertum var. clavatum	▬▬▬▬▬▬			
E. subarcticum	▬▬▬▬▬▬			
Eponides frigidus	▬▬▬▬▬▬	− − − − − −		
Globobulimina auriculata	− − − − − −			
Glomospira gordialis	− − − − − −	− − − − − −	− − − − − −	− − − − − −
Haplophragmoides bradyi				▬▬▬▬▬▬
H. glomeratum	− − − − − −		▬▬▬▬▬▬	▬▬▬▬▬▬
Hippocrepina indivisa			− − − − − −	− − − − − −
Hyperammina elongata		▬▬▬▬▬▬		
Labrospira crassimargo	− − − − − −	▬▬▬▬▬▬		
L. jeffreysii		▬▬▬▬▬▬		
L. cf. nitida				− − − − − −
Miliammina fusca	− − − − − −			
Nonion labradoricum	− − − − − −	− − − − − −		
Patellina corrugata	− − − − − −	−		
Proteonina atlantica				
Quinqueloculina arctica	− − − − − −			
Q. frigida	− − − − − −		− − −	
Q. seminula	− − − − − −	−		
Q. subrotunda	− − − − − −			
Recurvoides turbinatus	▬▬▬▬▬▬	− − − − − −		
Reophax arctica		▬▬▬▬▬▬	▬▬▬▬▬▬	▬▬▬▬▬▬
R. curtus		− − − − − −		
R. scottii		− − − − − −		− − − − − −
Spiroplectammina biformis	▬▬▬▬▬▬	− − − − − −		
S. typica	− − − − − −	− − − − − −	− − − − − −	− − − − − −
Triloculina tricarinata	− − − − − −	− − − − − −		
Trochammina advena				▬▬▬▬▬▬
T. inflata	− − − − − −			
T. lobata	▬▬▬▬▬▬	− − − − − −		
T. macrescens	−			
T. quadriloba			− − − − − −	
T. squamata				▬▬▬▬▬▬
Valvulina conica	− − − − − −	− − − − − −		
Virgulina complanata		− − − − − −		
V. fusiformis	− − − − − −	− − − − − −		
Textularia torquata	− − − − − −	− −		

Figure 31. Ranges of benthonic Foraminifera in the southern Gulf of Maine, off Portsmouth, New Hampshire. After Phleger (1952b).

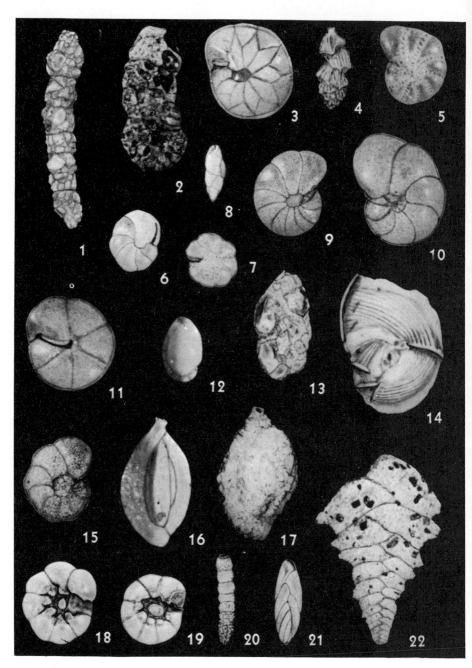

Plate 1. Benthonic Foraminifera occurring in the Gulf of Mexico at depths less than approximately 100 m.

EXPLANATION OF PLATE 1

1. *Bigenerina irregularis* Phleger and Parker. × 39.
2. *Ammoscalaria pseudospiralis* (Williamson). × 42.
3. *Asterigerina carinata* d'Orbigny. × 42.
4. *Angulogerina bella* Phleger and Parker. × 62.
5. *Elphidium poeyanum* (d'Orbigny). × 42.
6. *Epistominella vitrea* Parker. × 62.
7. *Buccella hannai* (Phleger and Parker). × 62.
8. *Buliminella* cf. *B. bassendorfensis* Cushman and Parker. × 70.
9. *Cibicides io* Cushman. × 42.
10. *Hanzawaia concentrica* (Cushman). × 42.
11. *Eponides antillarum* (d' Orbigny). × 42.
12. *Nonionella* cf. *N. opima* Cushman. × 70.
13. *Nouria polymorphinoides* Heron-Allen and Earland. × 42.
14. *Nodobaculariella cassis* (d'Orbigny). × 47.
15. *Planulina exorna* Phleger and Parker. × 42.
16. *Quinqueloculina compta* Cushman. × 48.
17. *Quinqueloculina horrida* Cushman. × 42.
18, 19. *Streblus beccarii* (Linné) variants. × 62.
20. *Rectobolivina advena* (Cushman). × 42.
21. *Virgulina punctata* Cushman. × 42.
22. *Textularia mayori* Cushman. × 33.

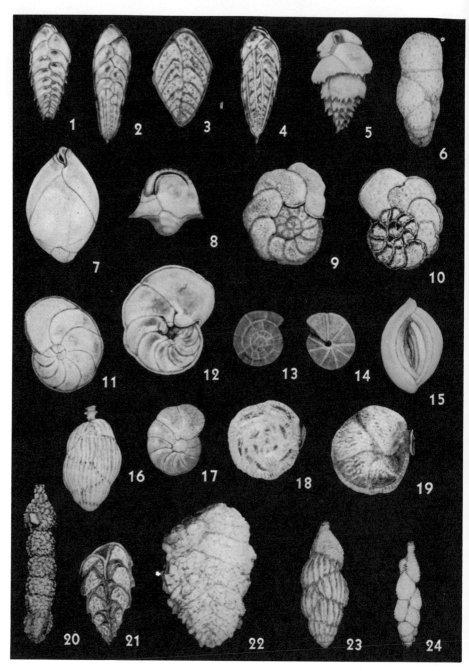

Plate 2. Benthonic Foraminifera occurring in the Gulf of Mexico at depths approximately 100-200 m.

EXPLANATION OF PLATE 2

         1. *Bolivina barbata* Phleger and Parker. × 62.
         2. *Bolivina fragilis* Phleger and Parker. × 60.
         3. *Bolivina goësii* Cushman. × 62.
         4. *Bolivina subaenariensis mexicana* Cushman. × 42.
         5. *Bulimina marginata* d'Orbigny. × 62.
         6. *Goësella mississippiensis* Parker. × 42.
         7. *Globobulimina mississippiensis* Parker. × 62.
         8. *Ehrenbergina spinea* Cushman. × 62.
9 , 10. *Planulina foveolata* (Brady). × 42.
11 , 12. *Rosalina bertheloti* d'Orbigny. × 42.
13 , 14. *Eponides regularis* Phleger and Parker. × 70.
     15. *Sigmoilina distorta* Phleger and Parker. × 62.
     16. *Uvigerina flintii* Cushman. × 42.
     17. *Cibicides deprimus* Phleger and Parker. × 42.
18 , 19. *Siphonina pulchra* Cushman. × 42.
     20. *Pseudoclavulina mexicana* (Cushman). × 24.
     21. *Reussella atlantica* Cushman. × 62.
     22. *Textularia foliacea occidentalis* Cushman. × 39.
     23. *Uvigerina parvula* Cushman. × 62.
     24. *Uvigerina laevis* Goës. × 62.

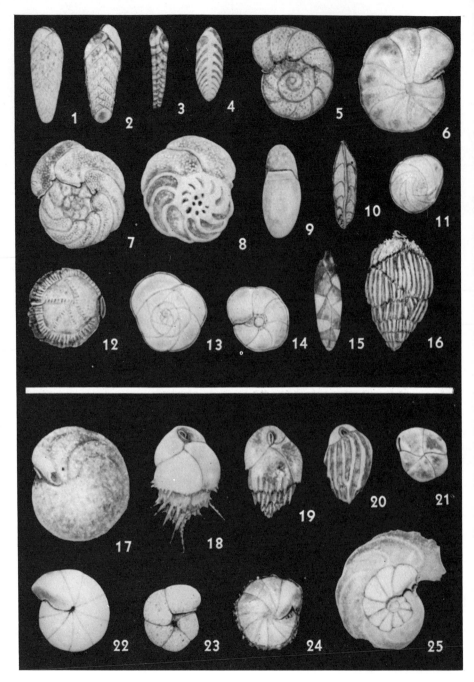

Plate 3. Benthonic Foraminifera from the Gulf of Mexico. Figures 1-16 are species which occur at depths approximately 200-500 m. Figures 17-25 are species which occur at depths greater than 500 m.

CHAPTER II 73

EXPLANATION OF PLATE 3

        1. *Bolivina albatrossi* Cushman. × 62.
        2. *Bolivina minima* Phleger and Parker. × 62.
        3. *Bolivina translucens* Phleger and Parker. × 62.
        4. *Bolivina ordinaria* Phleger and Parker. × 70.
5 , 6. *Cibicides umbonatus* Phleger and Parker. (5) × 42; (6) × 39.
7 , 8. *Planulina ariminensis* d'Orbigny. (7) × 30; (8) × 42.
        9. *Chilostomella oolina* Schwager. × 62.
      10. *Trifarina bradyi* Cushman. × 42.
      11. *Cassidulina curvata* Phleger and Parker. × 42.
      12. *Siphonina bradyana* Cushman. × 42.
13 , 14. "*Rotalia*" *translucens* Phleger and Parker. × 70.
      15. *Virgulina tessellata* Phleger and Parker. × 62.
      16. *Uvigerina hispido-costata* Cushman and Todd. × 42.
      17. *Cibicides rugosa* Phleger and Parker. × 30.
      18. *Bulimina aculeata* d'Orbigny. × 42.
      19. *Bulimina spicata* Phleger and Parker. × 62.
      20. *Bulimina alazanensis* Cushman. × 62.
      21. *Epistominella exigua* (Brady). × 62.
      22. *Gyroidina orbicularis* d'Orbigny. × 62.
      23. *Haplophragmoides bradyi* (Robertson). × 70
      24. *Osangularia cultur* (Parker and Jones). × 42.
      25. *Laticarinina pauperata* (Parker and Jones). × 30.

Plate 4. Benthonic Foraminifera from the Gulf of Mexico. Figures 1-16 are species which occur deeper than approximately 1000 m. Figures 17-25 are species which occur deeper than approximately 2000 m.

CHAPTER II 75

EXPLANATION OF PLATE 4

           1. *Plectina apicularis* (Cushman). × 42.
           2. *Ammoscalaria tenuimargo* (Brady). × 42.
           3. *"Ammobaculites"* sp. B. × 70.
4 , 5. *Eponides tumidulus* (Brady). × 94.
6 , 7. *Eponides turgidus* Phleger and Parker. × 70.
           8. *Adercotryma glomeratum* (Brady). × 42.
           9. *Gyroidina neosoldanii* Brotzen. × 42.
10 , 11. *Epistominella decorata* (Phleger and Parker). × 70.
      12. *Glomospira charoides* (Jones and Parker). × 62.
      13. *Trochammina globulosa* Cushman. × 42.
      14. *Pullenia bulloides* (d'Orbigny). × 42.
      15. *Siphotextularia curta* (Cushman). × 42.
      16. *Siphotextularia rolshauseni* Phleger and Parker. × 62.
17 , 22. *Cibicides kullenbergi* Parker. × 42.
      18. *Uvigerina auberiana* d'Orbigny. × 42.
      19. *Sigmoilina schlumbergeri* Silvestri. × 41.
      20. *Virgulina advena* Cushman. × 42.
21 , 25. *Cibicides wuellerstorfi* (Schwager). × 42.
      23. *Nonion pompilioides* (Fichtel and Moll). × 42.
      24. *Quinqueloculina venusta* Karrer. × 42.

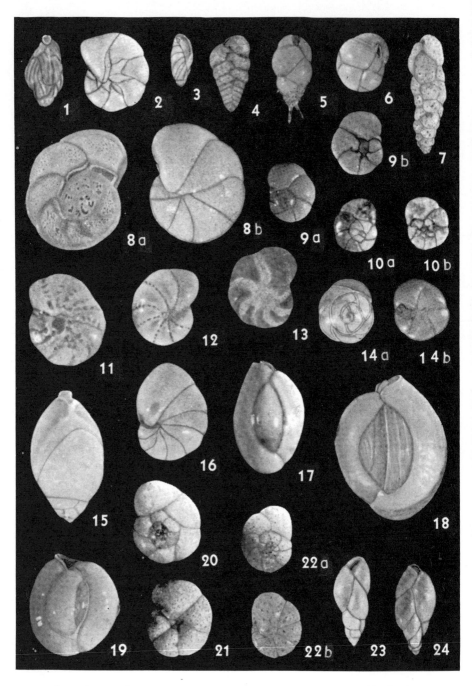

Plate 5. Benthonic Foraminifera from the southern Gulf of Maine which occur at depths less than approximately 60-75 m. ("sand facies").

EXPLANATION OF PLATE 5

      1. *Angulogerina angulosa* (Williamson). × 70.
      2. *Astrononion stellatum* Cushman and Edwards. × 70.
      3. *Buliminella elegantissima* (d'Orbigny). × 70.
      4. *Bolivina pseudoplicata* Heron-Allen and Earland. × 70.
      5. *Bulimina aculeata* d'Orbigny. × 48.
      6. *Cassidulina algida* Cushman. × 48.
      7. *Eggerella advena* (Cushman). × 48.
8a, b. *Cibicides lobatulus* (Walker and Jacob). × 48.
9a, b. *Rosalina columbiensis* (Cushman). × 48.
10a, b. *"Discorbis" squamata* Parker. × 70.
    11. *Elphidium incertum clavatum* Cushman. × 48.
    12. *Elphidium articulatum* d'Orbigny. × 48.
    13. *Elphidium subarcticum* Cushman. × 48.
14a, b. *Buccella frigida* (Cushman). × 70.
    15. *Globobulimina* (*Desinobulimina*) *auriculata* (Bailey). × 48.
    16. *Nonion labradoricum* (Dawson). × 48.
    17. *Quinqueloculina seminulum* (Linné). × 48.
    18. *Quinqueloculina arctica* Cushman. × 36.
    19. *Miliolinella subrotunda* (Montagu). × 48.
20, 21. *Trochammina inflata* (Montagu). (20) × 70; (21) × 48.
22a, b. *Trochammina lobata* Cushman. × 48.
    23. *Virgulina complanata* Egger. × 70.
    24. *Virgulina fusiformis* (Williamson). × 70.

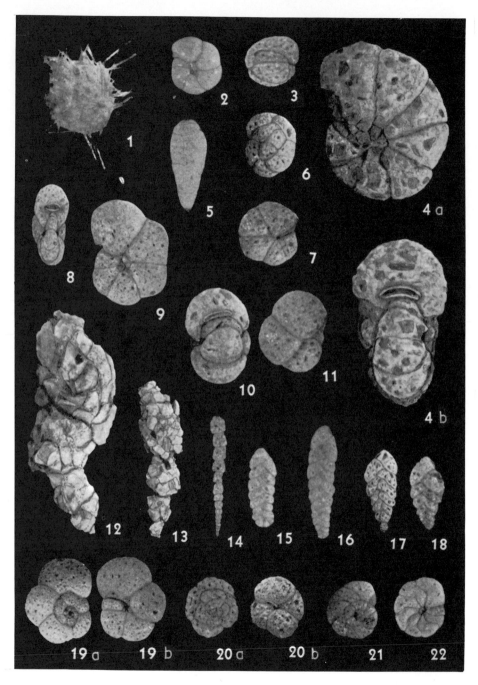

Plate 6. Benthonic Foraminifera from the southern Gulf of Maine which occur at depths greater than approximately 60-75 m. ("mud facies").

EXPLANATION OF PLATE 6

      1. *Crithionina pisum hispida* Flint. × 48.
      2. *Haplophragmoides bradyi* (Robertson). × 70.
      3. *Adercotryma glomerata* (Brady). × 48.
4a, b. *Alveolophragmium crassimargo* (Norman). × 48.
      5. *Hippocrepina indivisa* (Parker). × 48.
6 , 7. *Recurvoides turbinatus* (Brady). × 48.
8 , 9. *Alveolophragmium jeffreysii* (Williamson). × 48.
10 , 11. *Alveolophragmium* cf. *A.nitida* (Goës). × 48.
12 , 13. *Reophax curtus* Cushman variants. × 48.
     14. *Reophax scottii* Chaster. × 70.
15 , 16. *Spiroplectammina biformis* (Parker and Jones). × 70.
17 , 18. *Textularia torquata* Parker. × 70.
19a, b. *Trochammina advena* Cushman. × 48.
20a, b. *Trochammina quadriloba* Höglund. × 70.
21 , 22. *Trochammina squamata* Parker and Jones. × 70.

In this shelf area the forms most indicative of the inner turbulent zone down to 20-30 m. are *Ammobaculites dilatatus* Cushman and Brönnimann, *Buliminella elegantissima* (d'Orbigny), some species of *Elphidium*, *Gaudryina exilis* Cushman and Brönnimann, *Quinqueloculina compta* Cushman, *Q. lamarckiana* d'Orbigny, *Q. seminulum* (Linné), *Streblus beccarii* (Linné) in abundance, and *"Rotalia" rolshauseni* Cushman and Bermúdez.

Species apparently characteristic of the inner shelf shoaler than 50-70 m. in addition to the above forms are *Bolivina lowmani* Phleger and Parker, *B. striatula* Cushman, *Eggerella* sp., *Elphidium advena* (Cushman), *E. incertum mexicanum* Kornfeld, *E. matagordanum* (Kornfeld), *E. poeyanum* (d'Orbigny), *Nonionella atlantica* Cushman, *N. opima* Cushman, *Nouria polymorphinoides* Heron-Allen and Earland and *N.* sp.

Species restricted to the outer shelf deeper than 50-70 m., or more abundant there, and which may also occur deeper than the shelf are *Bigenerina* sp., *Bolivina fragilis* Phleger and Parker, *Cibicides* aff. *C. floridanus* (Cushman), *C. mollis* Phleger and Parker, *Eponides regularis* Phleger and Parker, *Gaudryina* cf. *G. aequa* Cushman, *Lenticulina peregrina* (Schwager), *Nonion formosum* (Seguenza), *Uvigerina parvula* Cushman and *Valvulineria minuta* Parker.

Species restricted to depths greater than approximately 90-100 m. are *Bulimina spicata* Phleger and Parker, *Cassidulina curvata* Phleger and Parker, *C. norcrossi australis* Phleger and Parker, *Pullenia quinqueloba* Reuss, *Sphaeroidina bulloides* d'Orbigny and *Uvigerina laevis* Goës.

Bandy (1956) has used median percentages of the total populations of significant Foraminifera to establish his continental shelf depth distributions in the eastern Gulf of Mexico. These data are used to delineate several shelf faunal zones as shown in Table 1.

TABLE 1. *Median percentages of significant Foraminifera in depth zones in the northeast Gulf of Mexico, based on concentrate samples. After Bandy (1956)*

Species	Depth zones (feet)					
	8–40	41–105	106–180	181–250	251–400	401–600
Fauna 1						
Elphidium gunteri	5	1	—	—	—	—
poeyanum	19	3	3	1	—	—
Streblus tepidus	48	1	1	—	—	—
Fauna 2						
Archaias angulatus	[1]43	11	1	—	—	—
Asterigerina carinata	5	20	1	1	—	—
Discorbis floridanus	2	8	1	1	—	—
concinnus	1	13	2	1	—	—
Hanzawaia concentrica	1	18	18	13	5	—
strattoni						
Textularia candeiana	—	4	5	3	—	—
mayori						
Fauna 3						
Bigenerina irregularis	1	4	10	10	4	1
Planulina ornata	—	2	16	16	5	—
Fauna 4						
Amphistegina lessonii	—	—	—	[2]1	[2]1	—
Cassidulina curvata assemblage	—	—	—	2	9	9
Cibicides pseudoungerianus	—	—	1	8	20	21
Gaudryina aequa	—	—	4	12	12	3
Textularia conica assemblage						
Fauna 5						
Bolivina goesii	—	—	—	—	1	7
daggarius	—	—	—	1	6	14
Planulina foveolata	—	—	—	—	1	2
Robulus calcar	—	—	—	1	1	3
Uvigerina bellula	—	—	—	—	6	18
flintii						
hispido-costata						

[1] Representative for areas of normal salinity only; 1 percent otherwise.
[2] Represents occurrence in concentrate; it would range from 10 to 70 percent of unconcentrated fauna.

Another technique for analyzing depth distributions is used by Walton (1955) who plotted the frequencies of the average numbers of living specimens taken at several collecting seasons against depth in populations in and near Todos Santos Bay, Baja California, Mexico. These ranges are shown in Figures 32 to 36. Walton's depth range data are among the most reliable yet published for the southern California and northern Mexico region since they are based on living assemblages taken seasonally. This eastern Pacific fauna differs somewhat from the Gulf of Mexico assemblages and those elsewhere. Walton's analysis of his distributions is listed below.

Living specimens of the following are restricted to depths less than 30 fms. (55 m.):

Quinqueloculina sp.
Elphidum translucens Natland
E. tumidum Natland
E. spinatum Cushman and Valentine
Dyocibicides biserialis Cushman and Valentine
Trochammina kellettae Thalmann
Bifarina hancocki Cushman and McCulloch
Textularia cf. *T. earlandi* Parker

Living specimens of the following are most abundant at depths less than 30 fms. (55 m.) but *occur* deeper:

Eggerella advena (Cushman)
Bolivina vaughani Natland
Cibicides fletcheri Galloway and Wissler
Buliminella elegantissima (d'Orbigny)
"*Rotalia*" spp.
Labrospira cf. *L. columbiensis* (Cushman) [=*Alveolophragmium*]
Discorbis spp. [=*Rosalina*]
Gaudryina cf. *G. subglabrata* Cushman and McCulloch
Reophax curtus Cushman
Cibicidina nitidula Bandy [=*Hanzawaia*]

(Text continued on page 88.)

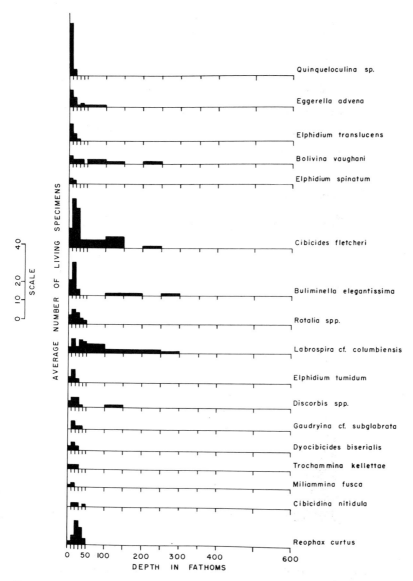

Figure 32. Frequency distributions of living benthonic Foraminifera in the Todos Santos Bay area, Baja California, Mexico. After Walton (1955).

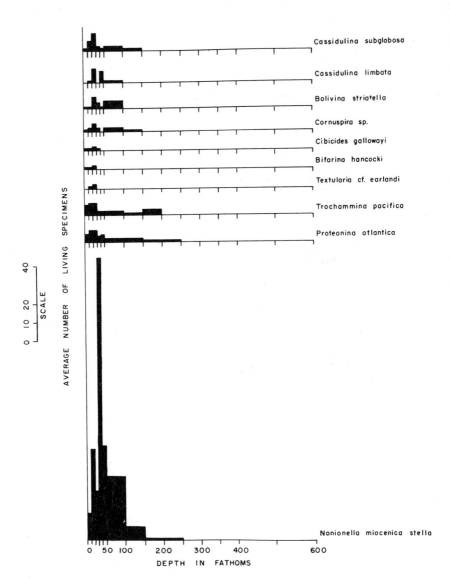

Figure 33. Frequency distributions of living benthonic Foraminifera in the Todos Santos Bay area, Baja California, Mexico. After Walton (1955).

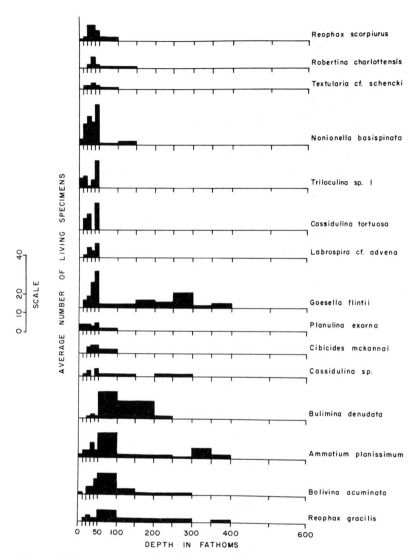

Figure 34. Frequency distributions of living benthonic Foraminifera in the Todos Santos Bay area, Baja California, Mexico. After Walton (1955).

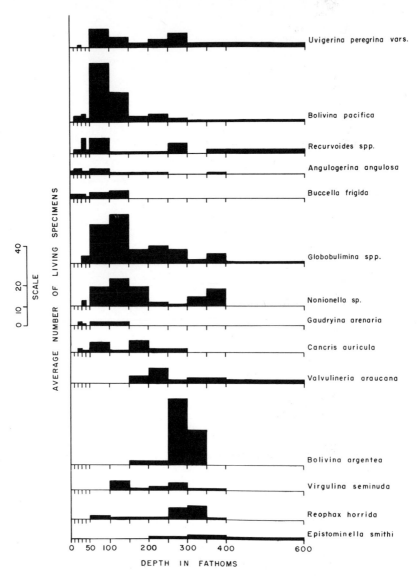

Figure 35. Frequency distributions of living benthonic Foraminifera in the Todos Santos Bay area, Baja California, Mexico. After Walton (1955).

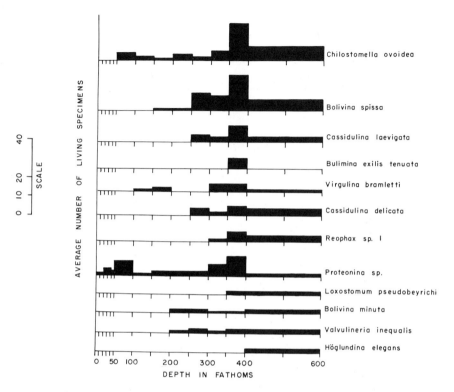

Figure 36. Frequency distributions of living benthonic Foraminifera in the Todos Santos Bay area, Baja California, Mexico. After Walton (1955).

88 Ecology and Distribution of Recent Foraminifera

Cassidulina subglobosa (H. B. Brady)
C. limbata Cushman and Hughes
Bolivina striatella Cushman
Cornuspira sp.
Cibicides gallowayi Cushman and Valentine
Trochammina pacifica Cushman
Proteonina atlantica Cushman

Living specimens of the following are most abundant at 30-50 fms. (55-91 m.):

Nonionella miocenica stella Cushman and Moyer
Reophax scorpiurus Montfort
Robertina charlottensis (Cushman)
Textularia cf. *T. schencki* Cushman and Valentine
Nonionella basispinata (Cushman and Moyer)
Triloculina sp. 1
Cassidulina tortuosa Cushman and Hughes
Labrospira cf. *L. advena* (Cushman) [=*Alveolophragmium*]
Goësella flintii Cushman
Planulina exornu Phleger and Parker
Cibicides mckannai Galloway and Wissler
Cassidulina sp.

Living specimens of the following *do not* occur deeper than 50 fms. (91 m.):

"*Rotalia*" spp.
Reophax curtus Cushman
Triloculina sp. 1
Cassidulina tortuosa Cushman and Hughes
Labrospira cf. *L. advena* (Cushman) [=*Alveolophragmium*]

Living specimens of the following are *most* abundant at 50-100 fms. (91-183 m.):

Bulimina denudata Cushman and Parker
Ammotium planissimum (Cushman)
Bolivina acuminata Natland
Reophax gracilis (Kiaer)

Uvigerina peregrina Cushman vars. (·part)
Bolivina pacifica Cushman and McCulloch
Recurvoides spp.
Globobulimina spp.
Nonionella sp. (part)
Angulogerina angulosa (Williamson)
Buccella frigida (Cushman)
Gaudryina arenaria Galloway and Wissler
Cancris auricula (Fichtel and Moll)

Living specimens of the following *do not* occur deeper than 100 fms. (183 m.):

Planulina exorna Phleger and Parker
Cibicides mckannai Galloway and Wissler
Textularia cf. *T. schencki* Cushman and Valentine
Reophax scorpiurus Montfort
Bolivina striatella Cushman
Cassidulina limbata Cushman and Hughes
Eggerella advena (Cushman)

Living specimens of the following are most abundant *between* 100 and 350-400 fms. (183 and 658-732 m.):

Valvulineria araucana (d'Orbigny)
Virgulina seminuda Natland
Reophax horrida Cushman
Bolivina argentea Cushman

Living specimens of the following in abundance are characteristic of water depths *greater* than 350-400 fms. (658-732 m.):

Chilostomella ovoidea Reuss
Bolivina spissa Cushman
B. minuta Natland
Cassidulina laevigata d'Orbigny
C. delicata Cushman
Bulimina exilis tenuata (Cushman) [=*Buliminella*]
Virgulina bramlettei Galloway and Morrey

Reophax sp. 1
Proteonina sp. (part)
Loxostomum pseudobeyrichi (Cushman)
Valvulineria inaequalis (d'Orbigny)

Höglundina elegans (d'Orbigny) generally does not occur shoaler than 400 fms. (732 m.).

Applications of Depth Range Information

DISPLACED FAUNAS

There are numerous records of sands in deep water in an environment apparently foreign to that of their original emplacement. An important source of information and interpretation of deep-sea sands in the San Diego Trough off California is a discussion by Ludwick (1950). One useful kind of evidence for the shallow-water origin of these sands and their displacement into deep water is the shallow-water benthonic Foraminifera which many of them contain. The basis for use of Foraminifera in determining the depth of origin of displaced sediments is the knowledge concerning depth ranges of the benthonic forms. This has been a useful tool in recognizing displaced sediments in the modern ocean and in ascertaining the probable origin and possible processes of displacement.

More than one kind of displacement has been suggested for the sands found in the San Diego Trough (Phleger, 1951a) where displacement of sediments has occurred over a distance up to 10 or 15 miles (see Figures 37-39 for analyses of these faunas). A coarse, shell sand occurred in core V 35 from the San Diego Trough at a depth of 1100 m. near the mouth of Coronado submarine canyon (Figure 37). The sand is similar to that present in shallow water in the San Diego area. The Foraminifera in this sand are all forms from depths

CORE V 35 1109 m.	TOP fecal pellets, quartz grains	-10" coarse shell sand	-32-36" fecal pellets	NORMAL DEPTH RANGES OF SPECIES IN METERS 200 400 600 800 1000
Angulogerina carinata		2(3)	.4(1)	————
Bolivina argentea	9(8)	6(12)	30(75)	
B. seminuda			.8(2)	————
B. spissa	10(9)		23(58)	
Bulimina spinifera	1(1)		2(4)	————————
B. subacuminata	1(1)		.8(2)	————————————
B. subaffinis	1(1)		1(3)	————————————
B. tenuata			1(3)	————
Cassidulina californica	3(3)	27(52)		————
C. laevigata	1(1)			————
C. limbata	20(18)	14(26)		———— - - - - -
C. quadrata	10(9)	.5(1)	2(4)	———————— - - - -
C. tortuosa	13(12)	32(61)	.8(2)	———— - - - - - - - - -
Chilostomella ovoidea	1(1)	.5(1)	11(29)	————
Discorbis cf. advena	2(2)			———
Elphidium crispum		1(2)		•
Epistominella smithi			4(9)	————————
Globobulimina pacifica	2(2)	.5(1)	7(17)	————————————
Gyroidina altiformis	1(1)	3(5)	.4(1)	
Höglundina elegans	2(2)	2(3)		————
Hopkinsina oceanica	2(2)			
Loxostomum pseudobeyrichi			2(6)	
Planulina exorna	1(1)			
P. sp.	4(4)	1(2)		———— - - - - - - - - - -
Pullenia salisburyi		.5(1)		- - - - - - - - - - - - - -
Robulus		4(7)		————
Uvigerina excellens	1(1)	1(2)		————
U. peregrina	8(7)	3(6)	11(28)	- - - - - - - -
U. senticosa		.5(1)		————
Valvulineria araucana			3(7)	————————————
Virgulina cf. bramletti			.4(1)	————————

Figure 37. Analysis of displaced faunas of benthonic Foraminifera in core V35 from the San Diego Trough. Redrawn after Phleger (1951a). First number, per cent of total population; number in parentheses, actual number of specimens.

less than approximately 300 m. and no deep-water species are present (*i.e.*, species from the depth at which the core was collected). Deep-water Foraminifera occur in the mud underlying and overlying the shell sand. This sand seems to have moved from shallow into deep water as a mass retaining its faunal and lithologic composition, as in a small landslide.

The silty sand at the top of core V 35 from San Diego Trough contains both shallow- and deep-water Foraminifera.

The upper part of the sediment containing the mixed fauna may have been a turbulent outer fringe of the displaced sediment. Population by deep-water species would occur at the sediment surface after deposition. A sand at the top of core V 54 (Figure 38), taken within ¼ mile of core V 35, also contains a fauna of shallow- and deep-water species. This may be a part of the same mass of sand.

Many of the other mixed faunas from the San Diego Trough probably were due to population with deep-water species after deposition of the sand in the new environment. The sand in core V 9 (Figure 39) on the east flank of San Diego Trough probably came from shallow water carrying with it specimens of shallow-water Foraminifera. Inclusion of the deep-water specimens could be accomplished either by physical mixing of the sediments and faunas or by population with the normal deep-water Foraminifera after deposition in its present position. The latter explanation seems more likely in the V 9 sand because it is well-sorted (So = 1.26) and undoubtedly came from a single source and depositing agency.

Some of the deep-water sands may have moved downslope to their present positions in more than one stage. A sand with such a history could be populated by Foraminifera at the sediment surface at each intermediate stop, provided the sand remained at the surface long enough for local Foraminifera to invade the material.

In the northwestern Gulf of Mexico the presence of displaced foraminiferal faunas is recorded from the Sigsbee Deep in the lower parts of cores 335 and 336 (Phleger, 1951a; shown in Figure 40). Shallow-water species occur in these cores at a depth of approximately 3500 m. associated with deep-water species characteristic of the depths in which the materials were collected. Some of the species are indigenous to depths of less than 100 m., and clearly indicate that the sediment must have originated in very shallow water. They suggest a displacement of at least 80 nautical miles and

CORE V54 1134 m. TOP COARSE QUARTZ SAND, SHELL FRAGMENTS		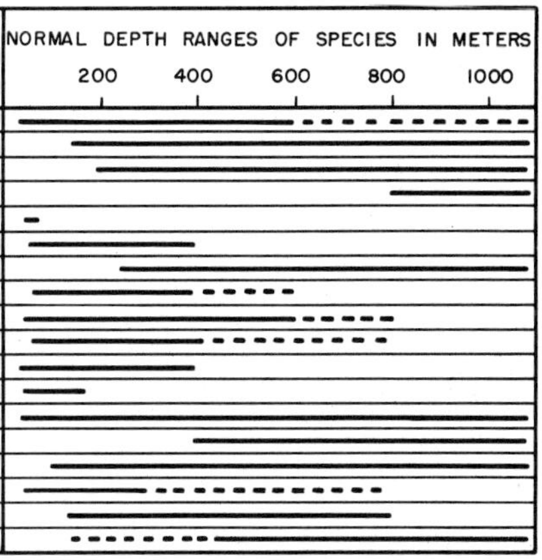 NORMAL DEPTH RANGES OF SPECIES IN METERS 200 400 600 800 1000
Angulogerina angulosa	2 (3)	----------- (800-1000)
Bolivina argentea	1 (2)	———— (200–400)
B. spissa	3 (5)	——— (200–400)
Bulimina subacuminata	1 (2)	——— (600–800)
Buliminella elegantissima	.7(1)	— (200)
Cassidulina californica	19 (28)	——— (200–400)
C. delicata	.7(1)	———————— (400–1000)
C. limbata	10 (15)	——— ····· (200–600)
C. quadrata	1 (2)	——— ····· (200–600)
C. tortuosa	46 (63)	——— ---------- (200–800)
Cibicides mckannai & vars.	3 (4)	——— (200–400)
Discorbis cf. advena	.7(1)	——— (200)
Globobulimina pacifica	.7(1)	————————— (200–1000)
Gyroidina altiformis	1 (2)	———————— (400–1000)
Höglundina elegans	3 (5)	——————— (200–800)
Planulina sp.	.7(1)	——— ------------ (200–800)
Robulus	1 (2)	———————— (200–1000)
Uvigerina peregrina	1 (2)	········——— (200–800)

Figure 38. Analysis of displaced faunas of benthonic Foraminifera in core V54 from the San Diego Trough. Redrawn after Phleger (1951a). First number, per cent of total population; number in parentheses, actual number of specimens.

CORE V9 744 m. -22" FINE SAND		NORMAL DEPTH RANGES OF SPECIES IN METERS 200 400 600 800 1000
Bolivina argentea	7 (4)	——————— (200–800)
B. minuta	2 (1)	
B. seminuda	7 (4)	——————— (400–800)
B. spissa	11 (6)	——— (200–400)
B. subadvena	6 (3)	——— (200–400)
Bulimina subaffinis	2 (1)	——————— (800–1000)
Buliminella elegantissima	2 (1)	— (200)
Cassidulina delicata	28 (15)	———————— (400–1000)
C. subglobosa	4 (2)	
Discorbis cf. advena	4 (2)	— (200)
Elphidium tumidum	2 (1)	— (200)
Epistominella smithi	4 (2)	——————— (400–800)
Eponides repandus	4 (2)	
Gyroidina altiformis	2 (1)	———————— (400–1000)
Höglundina elegans	2 (1)	——— (200–400)
Loxostomum pseudobeyrichi	2 (1)	———————— (400–1000)
Nonionella cf. atlantica	2 (1)	
Sigmoilina cf. miocenica	2 (1)	
Uvigerina excellens	7 (4)	——————— (400–800)

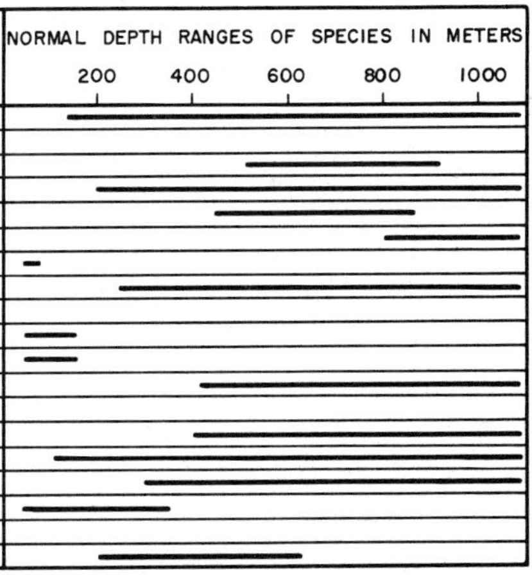

Figure 39. Analysis of displaced faunas of benthonic Foraminifera in core V9 from the San Diego Trough. Redrawn after Phleger (1951a). First number, per cent of total population; number in parentheses, actual number of specimens.

	CORE 335 - 3438 m.													CORE 336 - 3530 m.									
Cm. from top of core	0	24	56	105	138	156	168	178	182	186	193	198	214	1	28	41	96	117	133	153	168	189	204
Angulogerina bella								2			1	2								2			
Bigenerina irregularis											1								2				3
Bolivina alata										21										5	5	3	
B. albatrossi												2							4				
B. barbata											5	3								2		3	3
B. lowmani												8										5	
B. striatula spinata								5			3	2											
B. subaenariensis mexicana											3	5								5		11	
Bulimina aculeata												1							13	5		5	8
B. alazanensis											3	2							2				3
B. marginata								2			3	2									5		
B. striata mexicana								2			1	3										3	5
Cancris oblonga																			2	2			
Cassidulina subglobosa		1									1											5	
Cibicides aff. floridanus	3			3															4	5	2		5
C. wuellerstorfi	11	31	21	15	32	33			5			1	12	47	33	6	11	36			2	5	
Cibicidina strattoni											3	6							4	5	10	11	13
Eggerella bradyi	5		5	5	4	4							4			6	3	6					
Elphidium discoidale								5			31	24							18	25	20	8	30
E. cf. fimbriatulum												1							2				
E. gunteri											1	1							4	5		3	
Epistominella decorata	8	7	19	13	18	7			5		1		4		22	35	21	12	2				
Eponides polius	14	17	5											21									
E. tumidulus	3		9	5	8										11	12	14						
Globobulimina affinis						7	20	10	17	36								15	4	2	2	3	3
Glomospira charoides	25	1												5	11	6	3						
Gyroidinoides altiformis					4											12	3				2	3	
Nonion pompilioides	5	20	19	13	14	11	20						4	5	11		7	6					3
Nonionella atlantica												3							2	2	2		
Osangularia cultur		1	2								1												
Pseudoeponides umbonatus			5	10	4	7										6	3	9	2		2	3	
Rectobolivina advena											3	4									2		
Reussella atlantica								7		7	1								4	2			5
"Rotalia" beccarii parkinsoniana											8	9							4	7	7	3	3
"R." beccarii tepida																			2		2		
"R." translucens								2												2			3
Uvigerina parvula										21		2											
U. peregrina								10	5		5	6							11	2	12	11	5
Virgulina pontoni								5	11			2								2	2		

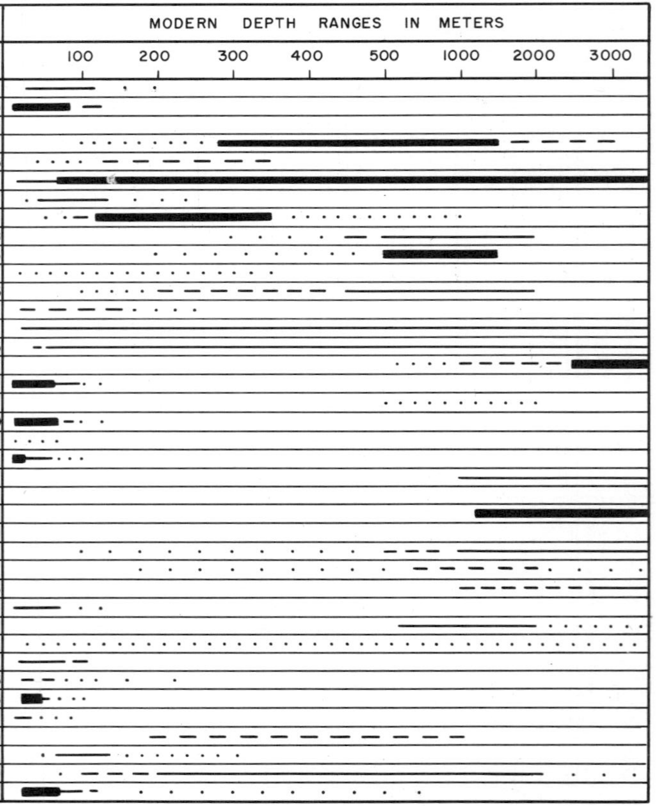

Figure 40. Analysis of displaced faunas of benthonic Foraminifera in two cores from the Sigsbee Deep in the Gulf of Mexico. After Phleger (1951a). Occurrences of species in per cent of total population for each sample.

through a depth range of more than 3000 m. The association with deep-water species indicates that the materials at their present location were deposited at depths in excess of 2000 m., the deepest reliable faunal boundary recognized in the area.

The displaced benthonic faunas in the two Sigsbee Deep cores suggest a complex history for the sands which contain them. The Foraminifera came from depths of from less than 100 m. to more than 2000 m. Faunal mixing could occur by downslope displacement in short stages, by mixing during a single movement, or by supply of sediment from several sources and depths. Core 336 is from the flat floor of the Sigsbee Deep, and more than 25 nautical miles from the base of the continental slope. Presence of sand containing shallow-water Foraminifera at this position indicates that sediment moving downslope may be carried onto a relatively flat basin beyond the base of the slope where it obtained its momentum.

The displaced faunas in cores 335 and 336 are correlatives and are of late glacial age, based upon the vertical sequence of planktonic Foraminifera (discussed in Chapter V). This indicates that the sands and displaced faunas probably were deposited by the same turbidity current. The dimensions of the sediment involved are approximately defined since the sand appears to cover an area at least 25 miles in one direction. There is no evidence of displaced sands in other cores nearby, suggesting that the sand does not cover an extensive area.

Greenman and LeBlanc (1956) have questioned the turbidity current origin of the "glacial" sands in these cores. They argue that the absence of graded bedding, from a megascopic examination of the cores, and presence of a sharp upper contact prove that they could not have been emplaced by this mechanism. Ludwick (1950) has shown, however, that not all the sand layers in the San Diego Trough have graded bedding, that some bedding can be recognized by microscopic examination, and that the upper contact of a sand layer may be either abrupt or gradational.

Greenman and LeBlanc agree that the sands in these two Gulf of Mexico cores are similar in appearance to shallow-water sand and that they contain both shallow- and deep-water Foraminifera. They suggest that these sands were deposited when the water was of depths similar to that of the continental shelf and that the Sigsbee Deep has been downfaulted in post-glacial time. The presence of deep-water Foraminifera is explained as representing a change in ecologic adjustment of these forms since glacial time.

Most of the other Sigsbee Deep cores which have been studied by the writer contain benthonic Foraminifera in the glacial sediments which are normal for the depth at which they occur, and the question arises of why other cores should not also contain mixed shallow and deep benthonic faunas if general downfaulting has occurred. Moreover, it seems highly improbable that there has been a change in the depth ranges of these deep-water forms since late Pleistocene. It is believed that the faunal and sedimentary evidence does not substantiate the hypothesis of extensive post-glacial downfaulting of the Gulf of Mexico basin.

Deep-water sands at some stations in the San Diego Trough contain only deep-water Foraminifera. These sands probably are of shallow-water origin since the principal sediment in the trough is mud and other deep-water sands nearby in the same area contain shallow-water species. The shallow-water specimens originally present may have been segregated from the sand in these samples by differential sorting of the sediment and it may have been devoid of Foraminifera when it was deposited. Specimens originally present may have been left behind during transport or carried elsewhere and deposited. These samples were collected from near the western side of the trough and at a considerable distance from a probable source of sand, adding circumstantial evidence for differential sorting.

Displaced foraminiferal faunas have been recorded from

the deep Atlantic (Phleger *et al.*, 1953). These are located near areas of relatively high submarine relief, as follows: near the Canary Islands, near the Cape Verde Islands, near Puerto Rico, and off the coast of northern South America. Off the coast of South America the displaced materials must have moved a distance of 300-400 miles and through a depth of more than 4000 m. Most of these records show that the displaced faunas occur in sands. There is one exception, however, off Puerto Rico, where the enclosing sediment is a *Globigerina* ooze containing a high per cent of finely-divided calcareous material which originated in shallow water.

Reports have been published on numerous long cores collected from the North Atlantic by Ericson *et al.* (1951, 1952). These authors record deep-water sands from a variety of areas and have generalized on certain aspects of the faunas as follows (Ericson *et al.*, 1951, p. 963): "Foraminifera, though never abundant, are usually present in the coarse sands. In addition to the usual planktonic species there are almost without exception a few species which are characteristic of the continental shelf and slope sediments, such as *Elphidium incertum, Globobulimina auriculata,* and *Nonion labradoricum.*" Ericson *et al.* (1952) list shallow-water species as occurring in seven of 81 cores and suggest faunal evidence for shallow-water origin of deep-water sands in a few others.

Early studies of Foraminifera of submarine cores from the North American continental slope (Phleger, 1939, 1942) show many faunas which contain shallow-water benthonic species displaced out of their normal depth range. This displacement was not recognized at the time the papers were published. F. L. Parker (1958) reports displacement of benthonic faunas from shallow into deeper water in seven samples from the eastern Mediterranean. All these samples were taken near the coast and near steep slopes leading up to shallow water. They contain a mixture of deep-water and shallow-water forms, demonstrating that they were deposited at the depth from which they were collected. In addition, she found displaced faunas in several cores taken in this area. Hamilton (1953, p. 216) has shown displacement of faunas from a flat-topped seamount to the floor of the deep Pacific basin. Natland and Kuenen (1951) have given convincing evidence for deposition of displaced faunas and sediment in deep basins during the Tertiary in the Ventura Basin, California.

The displaced faunas appear to be located near areas of considerable relief. Data from the cores collected by the Swedish Deep-Sea Expedition and others suggest that displaced faunas cover only a small percentage of an ocean basin, even though they may have considerable areal extent. It seems reasonable to assume that sediment being displaced to the deep ocean will carry an assortment of sediment from clay through sand sizes. Differential sorting of these sizes as the sediment is moved downslope and gradually loses velocity is expected, with the sand sizes deposited first nearest the source. Tests of Foraminifera may be considered as sand particles in their reaction to processes of sediment transportation and deposition so that any Foraminifera which are contained in the sediment will be deposited along with the sand-size material. They should be deposited along with grains of sand smaller than the tests themselves since Foraminifera have a relatively low density per unit volume. The fine silt and clay sizes of the material being displaced will be carried beyond the coarser material and deposited where there is no interference from a topographic barrier. The fine-grained sediment is not expected to carry tests of Foraminifera since they have been deposited with the sand.

Some or most of the fine-grained sediment in the deep North Atlantic, far from sources of supply, may be the fine fraction of displaced sediment. A considerable percentage of deep-sea sediment in the North Atlantic basins may be of this origin.

Studies by Ludwick (1950) and others in the San Diego Trough, off San Diego, California, have shown a relationship between displaced sediment and the presence of submarine canyons. It appears that most of the displaced material is funneled down the canyons in this area. Several cores taken from the continental slope off the Mississippi Delta have been studied (Phleger, 1955b). Two of these cores from the bottom of Mississippi Canyon contain microfaunas and other sediment which moved from depths less than approximately 100 m., and cores from the sides of the canyon as well as elsewhere on the continental slope do not contain such material. This is additional evidence that much sediment flows down the channels of submarine canyons and thus displaced sediment may be more prevalent in canyon areas than elsewhere. Ewing *et al.* (1958) conclude that a large cone of fine grained sediment off the Mississippi River was deposited by Pleistocene turbidity currents, presumably funneled there through the Mississippi Canyon. Their conclusions are based on sedimentary and topographic evidence.

CHANGE IN SEA-LEVEL

In the southern Gulf of Maine, off Portsmouth, New Hampshire (Phleger, 1952b), the shallow-water sand-facies fauna and the deeper-water mud-facies fauna apparently are real depth distributions (Figure 31). In 48 short cores examined from the mud facies there is a lower fauna, also in mud, composed almost entirely of calcareous benthonic Foraminifera. The fauna in the mud at the top of the cores is almost exclusively composed of arenaceous benthonic species. The lower-core calcareous fauna is essentially the same as the present-day sand-facies fauna which is restricted to shoaler water depths. The difference in water depth between the deepest occurrence of the modern sand-facies fauna and the deepest lower-core fauna is approximately 75-100 m.

This suggests that there has been a rise of sea-level since deposition of the calcareous fauna in the lower part of the cores. It appears that the redistribution of faunas was connected with rise of sea-level during post-glacial time. As the water deepened over the mud areas the environment may have become less desirable for the shallow-water calcareous species and they gradually invaded the newly-opened shoaler environment of the present sand facies.

It is generally believed that there was a lowering of sea-level during the glacial stages (see Flint, 1947, p. 437). This is based on the evidence for considerable thickness of glacial ice in North America and Europe and the fact that the water immobilized as ice would be subtracted from the ocean. It is estimated that sea-level must have been lowered 75-100 m., on the basis of indirect evidence as to the total volume of ice on the continents. The faunal data presented from the Portsmouth, New Hampshire, area appear to offer direct evidence concerning glacial sea-level lowering and post-glacial rise of sea-level, and also suggest a minimum lowering of approximately 75 m.

Three cores from the northwestern Gulf of Mexico, southeast of Galveston, Texas, are of interest in this connection (Phleger, 1951b, p. 78, table 35, fig. 32). These cores (462, 467, and 470) come from just below the shelf break at depths of 126 m., 157 m. and 229 m., respectively. The planktonic assemblages are the same throughout the cores and indicate that they were deposited during part or most of post-glacial time. They also contain benthonic species in lower sections of the cores which are adapted to depths less than 50 m., as follows:

> *Angulogerina bella* Phleger and Parker
> *Bigenerina irregularis* Phleger and Parker
> *Buliminella* cf. *B. bassendorfensis* Cushman and Parker
> *Elphidium discoidale* (d'Orbigny)
> *E.* cf. *E. fimbriatulum* (Cushman)
> *E. incertum mexicanum* Kornfeld
> *Nonionella atlantica* Cushman
> *Quinqueloculina lamarckiana* d'Orbigny
> *Streblus beccarii* (Linné)

These occurrences suggest that the sediment enclosing them was deposited when sea-level was lower, *i.e.*, in early post-glacial time.

The sequence of foraminiferal faunas in a core taken from the edge of the continental shelf, at a depth of 49 fm. (90 m.), in the northwest Gulf of Mexico also is of interest. This core was collected southeast of Matagorda Bay at 27° 52'N. Lat. and 95° 22.5'W. Long., and is approximately 7 m. long.

The upper 400 cm. of the core contain the mixed low-latitude and mid-latitude planktonic fauna identical with the modern Gulf of Mexico assemblage. At —425 cm. to —575 cm. the low-latitude planktonic Foraminifera are sparse and do not occur in some samples, and there is an increase in species which are adapted to water having lower surface temperatures than those in the modern Gulf of Mexico. This fauna, at —425 cm. to —575 cm., is interpreted as indicating colder surface water than at present exists in this area, but not as cold as indicated by some of the faunas reported in a previous study of cores from the continental slope and basin (Phleger, 1951b).

The benthonic fauna in the upper 575 cm. of the core is normal for the depth at which the core was collected. At —600 cm. there is a benthonic fauna which is at present characteristic of depths less than approximately 20 m. in the Gulf of Mexico.

This faunal sequence is interpreted as follows: The lower fauna from —600 cm. to —735 cm. represents the lowered sea-level of the last glacial or perhaps early post-glacial time. This indicates a lowering of sea-level of at least 40 fm. (73 m.). The fauna at —425 cm. to —575 cm. probably was deposited during early post-glacial time before the surface waters had warmed to their present temperatures. The upper 6 m., therefore, appears to represent post-glacial deposition.

Examination of the ranges of species in continental shelf traverses off central Texas (Figures 18-21) shows that the depth distribution of the living population is different from the depth distribution of the total population for some species. This is best illustrated by "*Elphidium* all spp." and *Streblus beccarii* var. A, which show living specimens confined to the inner shelf indicating that this is their present-day range. Abundant dead specimens of these species, on the other hand, also occur on the outer shelf. This part of the outer shelf fauna is interpreted as having been deposited when sea-level was much lower than at the present time. The fauna is mixed with assemblages normal for the present depth of water. Non-living specimens of shallow-water Mollusca also have been found on the outer continental shelf in the Gulf of Mexico (R. H. Parker, 1960). These faunas have been used by Curray (1960) to trace the post-glacial rise of sea-level across the shelf.

Walton (1955) has recognized a nearshore turbulent zone fauna on the outer shelf off Baja California, Mexico, indicating a lowering of sea-level of at least 25-30 fm. (46-55 m.). The presence of nearshore elements in a fauna from the outer continental shelf in the Bay of Bengal, shown to the writer by M. Poornachandra Rao, clearly indicates a former stand of sea-level at least 30 fm. (55 m.) lower than present. The "*Rotalia-Elphidiella*" assemblage illustrated by Houbolt (1957, photograph 23) is a very nearshore turbulent zone assemblage now occurring at depths greater than approximately 35 fm. (64 m.) in the central Persian Gulf. This clearly suggests a lowering of sea-level of 30 fm. (55 m.) or more in that area.

*Ecologic Factors Affecting
Depth Distributions of Foraminifera*

The causes of depth zonation and other distributions of Foraminifera are not clearly understood. The ecologic factors affecting distributions of marine organisms may be one, several, or all of the following: temperature, salinity, food supply, water chemistry, hydrostatic pressure, turbidity, turbulence, substrate, currents, biologic competition, disease, etc. At the present stage of our knowledge it is not possible to evaluate any one of these possible factors.

The effects of temperature and salinity on distributions of Foraminifera probably have been over-emphasized. This is due in part to the abundance of temperature and salinity data available, and the relative ease of making additional observations.

TEMPERATURE

Temperature is a general factor of great ecologic importance in marine and non-marine environments. In the ocean, temperature variation within the seasonal layer and also other bathymetric temperature distributions are of importance in affecting distribution of organisms. The effect of temperature on the geographic distributions of various marine invertebrates is discussed by Hutchins (1947, p. 326), who points out that ". . . most organisms have a given range of temperature over which survival is possible, and within the survival limits a somewhat narrower range of conditions over which reproduction and repopulation can be completed. Four critical levels can be recognized in these relationships, and are readily identified with the seasonal boundary conditions as in the following scheme:

>Minimum temperature for survival
>Minimum temperature for repopulation
>Maximum temperature for repopulation
>Maximum temperature for survival."

The general principles stated by Hutchins are expected to apply to bathymetric and other distributions of benthonic Foraminifera as well as to geographic distributions of the organisms which he studied. It is not intended to imply that temperature is the only important ecologic factor affecting distribution of Foraminifera, nor that it is even of critical importance in causing distributions of all species. Significant seasonal variations in water temperature occur only in the upper seasonal water layer, and while there are successively lower temperatures with increase in depth into the permanent thermocline, these are relatively constant with depth. It is apparent, however, that the species which live within any range of depth can tolerate the temperatures existing there. The relationship of depth assemblages of Foraminifera to generalized bathymetric temperature distributions is shown on Figure 41.

The collection of useful marine temperature data for ecological purposes requires an understanding of the oceanography of the area being investigated. Within the seasonal layer it is necessary to have a seasonal distribution of temperature observations, whereas in and below the permanent thermocline there is little significant seasonal variation. Many workers have placed undue importance on temperatures taken at the time and place of sampling. Mean temperatures are not necessarily the critical data, but actual temperature extremes and their durations also are needed. It is desirable to accumulate more specific information on temperature effects on Foraminifera to evaluate the importance of the various types of observations.

There are some experimental data on temperature effects on Foraminifera but these are suggestive rather than conclusive. According to Myers (1936, p. 12) "The optimum temperature for laboratory cultures of *Spirillina vivipara* at La Jolla, California, is 21°C. At 18°C. reproduction is considerably reduced, while at temperatures above 26°C. the reproduction becomes abnormal, usually resulting in degeneration of part or all of the developing young." In his culture of *Patellina corrugata,* Myers (1935b, p. 358) noted: "Under

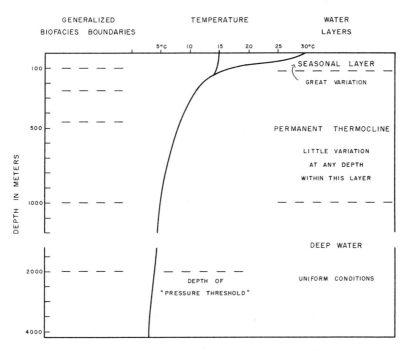

Figure 41. Depth faunas of benthonic Foraminifera related to generalized bathymetric distribution of temperature.

laboratory conditions the maximum rate of reproduction occurred at 21°C. Below 18°C. the rate of reproduction was greatly retarded. When a temperature of 25°C. was maintained for more than one day, reproduction became abnormal, and many of the larger individuals died. It is interesting to note that only 4 to 5°C. separate the optimum temperature from the upper thermal limit at which these species can be maintained in cultures." Cultures of *Discorbis, Pyrgo, Triloculina, Bulimina, Patellina, Spirillina* and *Robulus* survived at room temperature for more than a year but did not reproduce until the temperature was lowered to 18°C. or less (Myers, 1935a).

Bradshaw (1955) has made preliminary experiments on the effect of temperature on the rate of population increase in cultures of two species of *Rotaliella* from tide pools. He finds that at a temperature of 14.5°C. there is a much slower population increase than at mean room temperatures of 22.2°C. and 24.5°C. His populations were able to withstand maximum temperatures of 26°C. for short periods of time.

Arnold reports (personal communication) that he kept a culture of *Allogromia* alive through a temperature range from approximately 2°C. to 38°C. but with only short durations of the extreme temperatures. Arnold's material was collected from the shallow bay at Venice, Florida. From what is known of other, similar areas, it can be predicted that there exists a wide range of temperature in this bay. It is to be expected, therefore, that organisms which thrive in such an environment are able to withstand considerable environmental extremes.

Other temperature experiments by Bradshaw (1957) have been made on *Streblus beccarii* (Linné) in laboratory cultures. His results are figured graphically on Figure 42. There was no growth at temperatures below 10°C. or above 35°C. The specimens could survive the lower temperatures but at 35°C. and higher they died in a short time. Between these extremes there was growth with the rate increasing with higher temperature. Reproduction takes place between approximately 20-30°C. Temperature also affects the life span of each generation so that at 20°C. it may take four times as long for reproduction to occur as at 25-30°C. At 25-30°C. reproduction occurred approximately once a month. Forms grown at lower temperatures (<10°C.) grew to larger individuals than specimens grown at higher temperatures.

Streblus beccarii is one of the most abundant, typical forms of lagoons and nearshore areas where the environment may be characterized as being more variable than in most other

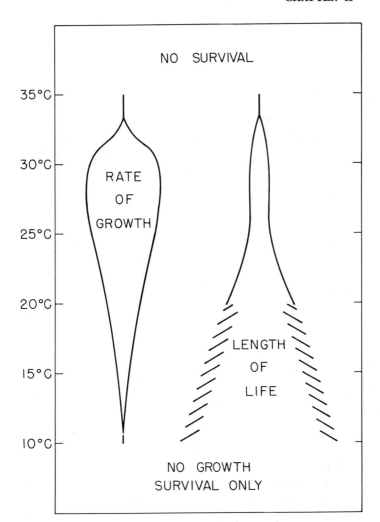

Figure 42. Generalized effects of temperature on rate of growth and length of life of *Streblus beccarii* (Linné) in laboratory cultures, based on work by Bradshaw.

places. It is possible that these experimental data may be of general application to all species of the fauna which lives in this environment.

These meagre experimental data on temperature effects suggest the following:

1. Temperature is effective in controlling the viability and repopulation ability of some Foraminifera.

2. Certain species of Foraminifera are able to withstand extreme temperature ranges, especially those forms which are adapted to environments where extremes are normal.

3. Different species have different temperature tolerances.

It should be pointed out that these experimental data are only on species from the littoral and lagoon areas, and the temperatures involved do not necessarily apply to species in other environments and may not apply to other species from the same environment.

SALINITY

Much has been written about the possible effect of salinity on the distribution of Foraminifera. Z. M. Arnold has stated (personal communication) that his cultures of *Allogromia* will withstand considerable range in salinity. This species comes from an environment where great ranges in salinity are common and such a tolerance is to be expected. Salinity variations probably do not aid in interpreting depth assemblages since it appears that they are too slight to be ecologically significant with depth in offshore water. Where salinity variations are marked, as in nearshore areas of high runoff, this factor may be of considerable importance although it is not known that salinity itself is a critical factor. It may be assumed, for example, that the fauna of marine marshes is adapted to the low salinity which usually occurs; but in some marine marshes there are relatively high salinities much of the time.

Bradshaw (1955) has shown experimentally in cultures of *Rotaliella heterocaryotica* Grell that the optimum population growth occurs at salinities of 26-30 o/oo. He found no significant growth in the population below 25 o/oo or above 37 o/oo, and no growth at salinities of 16.8-20.1 o/oo.

In laboratory cultures of *Streblus beccarii*, Bradshaw (1957) found that salinities between 20 o/oo and 40 o/oo allow normal growth and reproduction. Above and below these limits growth and reproduction activity fell off until growth finally ceased at salinities higher than 67 o/oo or lower than 7 o/oo. Reproduction occurred only at salinities of 13 o/oo to 40 o/oo. Generation time also was increased as the extreme tolerance limits were approached so that it required twice as long for reproduction to occur at 13 o/oo as it did within the normal range of salinity (20 o/oo-40 o/oo). Bradshaw's results on this species are illustrated graphically on Figure 43.

Salinity is a useful guide in distinguishing different types of marine water which may have different ecologic effects. This is best shown in some lagoons where mixing of river water with marine water results in a water of intermediate salinity. The distinctive lagoon foraminiferal assemblage is adapted to this environment, and even though salinity may not be the primary ecologic factor it is at least a guide to the type of water present.

Along coasts where there is appreciable runoff a wedge of nearshore water, as distinct from offshore water, may extend outwards from shore (Figure 4). This nearshore water can be recognized by having a lower salinity than offshore water. In the northern Gulf of Mexico such inshore water often extends to the bottom to a depth of approximately 50-60 m. on the continental shelf. This depth is a recognized faunal boundary in that area and it seems likely that the inner-shelf benthonic fauna is adapted to the nearshore, low-salinity water while the outer-shelf fauna is adapted to more oceanic water.

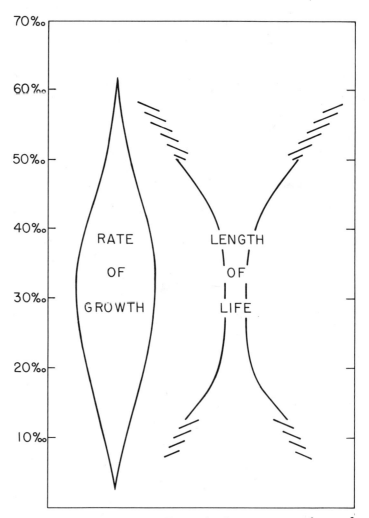

Figure 43. Generalized effects of salinity on rate of growth and length of life of *Streblus beccarii* (Linné) in laboratory cultures, based on work by Bradshaw.

FOOD

The quantity and kind of food are of primary importance in affecting distributions of all types of animals. Some animals will take only certain types of food and are therefore restricted to specific associations. The principal food of Foraminifera appears to be diatoms and flagellates such as *Nitzschia, Navicula,* and other diatoms (Myers, 1937, p. 94). According to Myers (personal communication), the Foraminifera which he studied seemed to accept any particulate food. It is to be expected that the largest production of Foraminifera would occur where there is the greatest amount of organic production and consequent supply of food. It is probable that kind and amount of available food are critical factors in distribution, rate of production, and rate of growth of some or most species. Boltovskoy (1954a) has attributed abnormalities in foraminiferal tests on the continental shelf off Argentina to unfavorable food conditions.

Culture experiments by Bradshaw (1955) have demonstrated the importance of food. It has been noted, for example, that too much food, in the form of flagellates or diatoms, has a deleterious effect, and too little food obviously has a bad effect. Certain forms seem to prefer flagellates which have been killed, but others will eat only living forms. Some species will eat only flagellates, some appear to eat anything, and one form is believed to eat only one species of diatom.

CHEMICAL EFFECTS

It is known that certain marine plants and animals require various chemical elements in small quantities for their development. A trace of soluble iron-bearing compound, for example, is necessary for culturing many microscopic aquatic plants. Distributions of trace elements in the ocean are not well-known, and it is possible that different marine water types contain different amounts of these materials. The ratios of the major chemical constituents of sea water are expected to be quite uniform in offshore areas. In nearshore areas where there is dilution of marine water by land runoff, or where concentrations of salts occur in evaporation basins, it is presumed that the over-all chemistry of the water is quite different from that in the open ocean.

Oxygen deficiency and/or presence of hydrogen sulfide are well-known factors in limiting distributions of other organisms, and it may be assumed that they affect the Foraminifera.

Hydrogen-ion concentration is a possible factor of importance although nothing is known about the experimental effects of pH on Foraminifera, and relatively few useful data are available on values in either bottom water or surface sediments. Some pH data in lagoon and marsh sediments are discussed in Chapter III. Hydrogen-ion concentration is expected to affect the production of calcareous tests so that at a pH less than approximately 7 such forms may not survive. This may not be a critical environmental factor in depth distribution of faunas since the pH of open-ocean surface water ranges from approximately 7.9 to 8.3 (Harvey, 1955, p. 55). Krumbein and Garrels (1952, p. 3) give the pH of the normal marine open-circulation environment as 7.8 at the bottom; presumably this is the bottom sediment. ZoBell (1946) gives the pH of most presumed open-ocean sediments as ranging from 7.5 to 9.5. According to ZoBell, "sandy bottom deposits" along the coast have a pH of 8.0 to 8.3, and this is probably the pH of the surrounding water. There is very little known concerning the range in pH of sediments with depth. The values vary somewhat depending upon the nature of the sediments, generally being lower in fine-grained materials than in coarse-grained ones.

Boltovskoy (1956) has commented on the "somewhat depauperate fauna" of Foraminifera on the continental shelf off Argentina. This is seen in small number of specimens, small size of individuals, loss of ornamentation, tendency toward asymmetry, and retarded growth. He suggests that the presence of lead, observed in analyses of the Foraminifera tests, has been the cause of this relatively poor faunal development by decreasing the productivity of marine plant life in this area.

TURBIDITY

Stainforth (1952, p. 43) reports that in interpreting the abundance of arenaceous Foraminifera in various ancient sedimentary facies in Trinidad, "our tentative conclusion was that turbidity was the controlling factor. The reasoning behind this conclusion was twofold: first, turbidity is the most obvious factor common to the three facies studied but unimportant in normal marine facies where arenaceous and calcareous Foraminifera coexist; second, photosynthesis is vital to the existence of calcareous organisms and is reduced or eliminated in areas of highly turbid water." There is no evidence which would suggest that photosynthesis is vital to the existence of either calcareous or arenaceous benthonic Foraminifera. Light is necessary for photosynthesis in marine plants, and can only affect marine animals insofar as the plants are essential in the food chain. Turbidity may have some effect on foraminiferal distributions but the writer is not aware of any direct evidence in this connection.

SUBSTRATE

Some Foraminifera require a foundation on which to attach and some forms seem to be always associated with "calcareous bioherms" or "coral reefs." There is no certain evidence that the material and nature of the substrate are of primary ecologic importance except in the two associations indicated. Where good correlation between faunas and sediments occurs, the correlation often seems to be a fortuitous one. This may be illustrated by the nearshore facies described from the southern Gulf of Maine, off Portsmouth, New Hampshire (Phleger, 1952b). In this area the sediment is primarily sand wherever the water depth is less than 60-80 m., and contains a foraminiferal fauna which is almost exclusively of calcareous species. In deeper water the sediment is composed of silt and clay and the Foraminifera are exclusively of arenaceous species. The correlation between the faunal assemblages and the sediments is so good that they are called a "sand facies" and a "mud facies" for descriptive purposes.

The clue to interpreting the sand and mud facies in the Portsmouth area is in comparing the distributions of the species with their distributions in nearby areas. Records of the Portsmouth species which occur in the Barnstable, Massachusetts, area (Phleger and Walton, 1950), and in the Long Island Sound–Buzzards Bay area (Parker, 1952) show that the same species in these other areas are not restricted to particular types of bottom. Occurrence of sand facies species in the mud in the lower parts of short cores from the Portsmouth area is further evidence, and it is probable that sediment type has little or no effect upon distribution of most of these Foraminifera. Several species restricted to the deeper, mud facies have a similar depth restriction on the outer continental shelf south of Cape Cod (Parker, 1948, fig. 2). Both the sediment and the foraminiferal faunas appear to be related to water depth and correlation between faunas and sediments is fortuitous.

Additional evidence on the relationship between foraminiferal assemblages and sediments is from traverses taken in the northwestern Gulf of Mexico. Stetson's (1953, pp. 7-19) data show that the sediment on the continental shelf is quite variable, or patchy, in type and distribution; the sediment on the continental slope and in the Sigsbee Deep (3500 m.) is relatively uniform throughout the entire area extending from the Mississippi Delta to the International Border.

Figure 44 shows the median diameters and the sorting coefficients of sediment along three of the traverses from which the foraminiferal faunas were taken. The generalized depth assemblages of Foraminifera are shown at the top of Figure 44. The lower traverse is off Port Isabel out to the center of the Sigsbee Deep. The main features of this traverse are: (1) the sediments tend to be coarser on the continental shelf than on the slope and basin but the median is quite variable and the sediments beyond the shelf are finer and more uniform, and (2) the sorting of the shelf sediments is very erratic, with some very poorly sorted, while the sorting of the slope and basin materials is more consistent and is generally better. The middle traverse shows an essentially similar distribution of sedimentary parameters off the Brazos River. In the traverse off Ship Shoal there are fewer differences between sediments at various depths, except that the slope and basin sediments are slightly finer-grained.

How do the distributions of the sediment parameters correlate with the distributions of the faunas? Obviously there is a difference between the shelf sediments and the rest of the sediments along these traverses. Many of the shelf sediments are generally coarser than the rest, and they are much more variable. This relative coarseness may be said in a general way to correspond to distributions of the shelf faunas at depths shoaler than 100 m. There is no apparent correlation, however, with any of the other depth distributions on the shelf (not shown on Figure 44). What does the very general correlation of the shelf sediment with assemblage 1 really signify? The coarseness and variation of this shelf sediment may be largely due to the shallow nature of this area and seasonal variations in turbulence. Processes of transportation and deposition are expected to be much more variable on the shelf than in deeper water. Also, strong winds causing intensive water turbulence will occasionally agitate the sediments on the bottom. In addition, as pointed out by Stetson

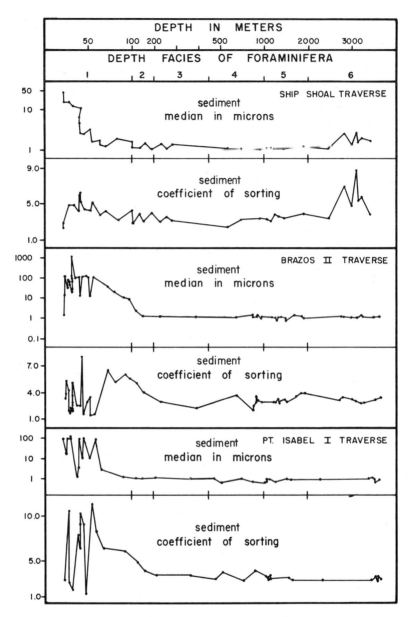

Figure 44. Relationship between depth assemblages of Foraminifera and mechanical sorting parameters of the sediment in three traverses in the northwestern Gulf of Mexico.

(1953), there is a likelihood that some of these sediments were deposited during a previous cycle, during lowered sea-level of a glacial stage. Such correlation as exists between the sediments and the faunas—and it is very general and very poor—seems to be coincidental.

There are rather striking faunal ranges at depths greater than 100 m., as shown in the figures illustrating depth distributions of the species in the northwestern Gulf of Mexico (Figures 14-17). These assemblages are summarized at the top of Figure 44. However, the sediment at these depths is essentially uniform, at least insofar as the parameters of median and sorting indicate. The different assemblages at these depths, then, are obviously developed regardless of the physical parameters of sediment illustrated.

Bandy (1954, p. 135) in his study of nearshore Foraminifera from the northern Gulf of Mexico reports that "there is no obvious consistent correlation between median grain size and faunal trends."

The depth assemblages reported by F. L. Parker (1948) from the Atlantic continental shelf and upper slope shown in Figures 29 and 30 are instructive in this connection. She found that the fauna is rather uniform between 15 m. and 90 m. in this region. Stetson's (1949) report of the sediment from the same samples shows considerable variation in the sediment type.

From the information available it appears that bottom sediment type is not a limiting factor in distribution of most benthonic Foraminifera, except where certain forms are attached to a hard bottom and in the coral reef assemblage. It is quite probable, nevertheless, that sediment type does have some ecologic effect, either directly or indirectly, on the distribution of benthonic Foraminifera. Fine-grained sediments generally contain a larger amount of organic material, and thus more potential food, than coarse-grained sediments (Trask, 1953). Fine-grained sediments may support a larger population than clean sands, especially where there is a mixture of mud and sand. Another example of apparent relationship recently has been found in some lagoons along the southwest coast of Texas and of California. In these areas some species of miliolids are mostly restricted to areas of sand or shell sediment. Numerous specimens of this group which have been preserved in formalin have been observed with pseudopods still attached to grains of sand or shell. In addition, the sands often support very small living populations.

An example of a form related to bottom type is seen in the distribution of *Cibicides lobatulus* (Walker and Jacob) in the southern Gulf of Maine off Portsmouth, New Hampshire, shown in Figure 45. This species is an attached form which has a discrete distribution in the area. It is confined to areas of hard bottom such as rocky, stony, and firm sand bottom, and its occurrence shown on Figure 45 can be correlated directly with the distribution of these types of substrate.

F. L. Parker (1952, p. 440) reports that the following attached species occur on sandy or stony bottoms in the Long Island Sound—Buzzards Bay area:

> *Cibicides lobatulus* (Walker and Jacob)
> *Discorbis columbiensis* Cushman [=*Rosalina*]
> *Poroeponides lateralis* (Terquem)
> *Quinqueloculina subrotunda* (Montagu) [=*Miliolinella*]

BIOLOGIC EFFECTS

There are biologic-ecologic factors which are difficult or impossible to evaluate at the present time. A community of benthonic Foraminifera may be thought of as a series of overlapping populations consisting of species, subspecies ("varieties"), and "races." When frequency distributions of species were plotted in the area off Portsmouth, New Hampshire (Phleger, 1952b), some of the more abundant species showed centers of high frequency with decreasing frequencies in all

CHAPTER II   119

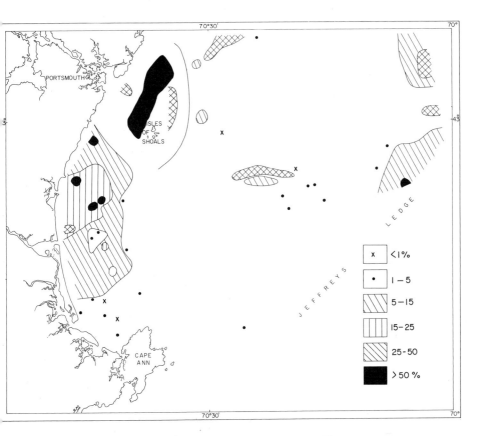

Figure 45. Distribution of *Cibicides lobatulus* (Walker and Jacob) off Portsmouth, New Hampshire. After Phleger (1952b).

directions from the centers (Figure 46). This suggests a population pattern which may be characteristic of benthonic Foraminifera. Foraminifera undoubtedly are subjected to rather intense intra-specific and extra-specific biologic competition. During times of stress, such as limited food supply or other borderline conditions within the environment, some species may be displaced from the area or reduced in abundance. It is also possible that disease may be important in limiting certain distributions.

EFFECT OF THE SEASONAL LAYER ON DEPTH DISTRIBUTION

In the northwestern Gulf of Mexico the major faunal boundary occurs at approximately 100 m. and varies from approximately 80 m. to 120 m. This coincides with the approximate depth of the seasonal water layer, and this undoubtedly is the most marked depth boundary occurring in many or most areas. From an ecological point of view this depth is a boundary between water having many variable properties, the seasonal layer, and water having relatively constant properties, the permanent thermocline (see Figure 41).

Some of the important characteristics of the seasonal layer which distinguish it from deeper water are:

1. Marked seasonal temperature range in mid-latitudes. Lowered salinity may occur in some coastal areas.

2. Effective light penetration for photosynthesis, thus the greatest production of organic matter by marine plants.

3. Turbulence during times of strong surface winds. This causes water mixing which will agitate the bottom materials and result in turbidity especially in shoal water. Water mixing and overturn also cause replenishment of nutrients to upper layers which increases production of organic matter.

The main depth boundary between foraminiferal assemblages in the northwestern Gulf of Mexico, 100 m. ± 20 m.,

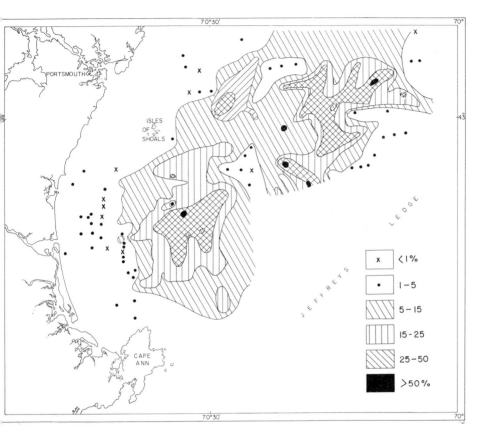

Figure 46. Distribution of *Reophax curtus* Cushman off Portsmouth, New Hampshire, showing frequencies decreasing from centers of high population. After Phleger (1952b).

is comparable to the 90 m. boundary proposed by F. L. Parker (1948) for the continental shelf off the eastern United States. Parker's 90 m. boundary also coincides with the approximate average depth of the seasonal effect. In the southern Gulf of Maine (Phleger, 1952b) the equivalent boundary is somewhat shoaler, at 60-75 m. This reflects the somewhat shoaler summer warming of the water in this part of the Gulf of Maine when there is a strong thermocline developed in the upper 40-50 m. of water and some warming to a depth of approximately 75 m. The boundary between "nearshore" and "offshore" depth zones off Barnstable, Massachusetts (Phleger and Walton, 1950), probably reflects the zone of intensive turbulence in shoal water.

Said's (1950) faunal depth boundary at 70 m. in the Red Sea can be correlated with the reported water-mass distribution in the area. According to Thompson (1939) there is active mixing to a depth of 80 m., which agrees approximately with Said's 70 m. boundary. The boundary at approximately 50-60 m. off southern California, first reported by Natland (1933) and also shown off Baja California by Walton (1955), seems to reflect hydrographic conditions in the upper layers of water along that coast.

INTERPRETATION OF OTHER DEPTH BOUNDARIES

Examination of Figure 41 shows additional faunal depth boundaries which have been recognized in various areas at the approximate depths of 200-300 m., 500 m., 1000 m., and 2000 m. It is difficult to interpret these deeper zones. It may be presumed that the boundary at approximately 1000 m. represents the bottom of the permanent thermocline. Below this the general water characteristics are more or less uniform, and above this depth there are at least differences in temperature and probably also other environmental factors.

The depth zones which occur within the permanent thermocline may represent adjustments of various species to different temperatures or temperature ranges, or to other factors not yet known.

The boundary at approximately 2000 m. does not represent any significant temperature, salinity, or other obvious difference in the water environment. Work by ZoBell and his colleagues on the effect of pressure on the growth of marine bacteria is of interest in this connection, and the following discussion by Oppenheimer and ZoBell (1952, p. 16) is pertinent:

"Although our knowledge of the effects of pressure on organisms is still woefully scant, . . . pressures such as those that occur in the sea may have multiple physiological effects. Slightly increased pressures may be stimulatory, whereas higher pressures cause the attenuation or death of the organisms. . . . By virtue of its effect on solubilities and dissociation constants, the pH and Eh [oxydation-reduction potential] values of complex solutions may be affected by pressure. These and other properties of solutions may combine to influence the physiology and ecology of marine organisms in many ways.

". . . The results presented in this paper substantiate the observations of ZoBell and Johnson (1949) that the pressure tolerance of marine bacteria is related to their depth habitat. The intolerance of certain bacteria for pressures exceeding 200 atm., for example, suggests that such organisms are not active in the sea at depths exceeding 2000 m."

This is the only specific evidence known to the writer concerning the effect of pressure on marine organisms. Hydrostatic pressure is an obvious ecologic factor which varies directly with depth in the ocean. It is possible that the boundary at 2000 m. can be explained as a "pressure threshold" at 200 atmospheres pressure as suggested by the results of Oppenheimer and ZoBell. Other depth boundaries also may be affected by hydrostatic pressure, either directly or indirectly.

CHAPTER III

Marginal Marine Distributions of Benthonic Foraminifera

In studying the environment of deposition of the older marine rocks the recognition of ancient shorelines is of considerable interest and importance. It is desirable, moreover, to designate the type of shoreline insofar as this is possible. Marginal marine areas are defined here to include marine marshes, estuaries, lagoons and bays, beaches, and the adjacent part of the continental shelf. These environments all have one thing in common—they are more affected by the land than are the environments farther offshore. Land runoff mostly affects these environments; much, or most, of the land-derived sediment is deposited here; many of the areas have turbulent water; there are marked variations in temperature; and there are great variations in other environmental factors.

In recent years some insight has been gained about certain marginal marine faunal distributions which do not appear to be a direct function of depth of water. The faunal patterns in these areas appear to be primarily affected by runoff from the land, by the conformation of the coastline, and by the consequent distribution of marine and semi-marine water masses.

Distribution of Assemblages in Gulf of Mexico and Northeastern United States

BARNSTABLE HARBOR, MASSACHUSETTS

Barnstable Harbor is on the southern shore of Cape Cod Bay in eastern Massachusetts. It is largely a marine marsh area with tidal channels and is separated from the open ocean of Cape Cod Bay by a sand spit approximately 6 mi. long and ½ mi. wide. The harbor is connected with the bay by a tidal channel about 400 yds. wide and there is a tidal range of several feet. There is relatively little runoff into Barnstable Harbor. Study of this area (Phleger and Walton, 1950) has shown the presence of one foraminiferal fauna characteristic of the bay and another restricted to the harbor area.

The Barnstable Harbor assemblage is characterized by abundance of *Trochammina inflata* (Montagu); the open-ocean fauna contains *Proteonina atlantica* Cushman and *Eggerella advena* (Cushman) in abundance. Three "subfacies" in Barnstable Harbor occur in the high marsh, the intertidal flats, and the channels. These are related to tidal action, nature and movement of bottom materials, presence of marsh grass, and relative organic production. The largest population occurs in the high marsh.

Figure 47 shows the geography of the Barnstable Harbor area and patterns of the foraminiferal faunas ("facies").

LONG ISLAND SOUND

The distribution of benthonic faunas of Foraminifera in the Long Island Sound–Buzzards Bay area is reported by

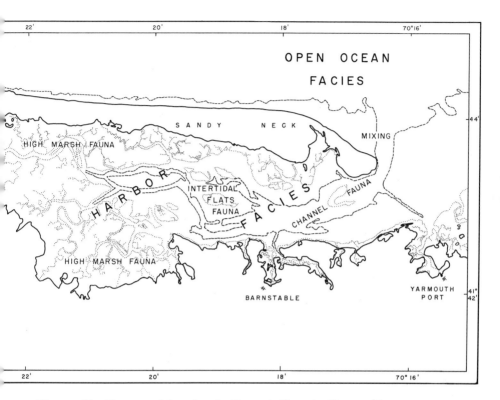

Figure 47. Faunas of benthonic Foraminifera in Barnstable area, Massachusetts. After Phleger and Walton (1950).

F. L. Parker (1952). The area studied is along the coasts of Connecticut and Massachusetts. There are "bays" and "sounds" which are partially separated from the open Atlantic Ocean by barrier islands. These islands do not form a continuous chain, but are placed so that the inland waters have different degrees of protection from the open ocean (see Figure 48). Long Island Sound, on the west, is effectively separated from the Atlantic by Long Island, except for a relatively narrow

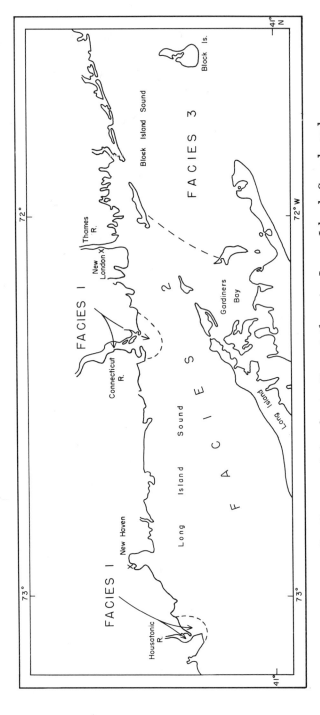

Figure 48. Faunas of benthonic Foraminifera in Long Island Sound and Block Island Sound. After F. L. Parker (1952).

passage on the east. Three large rivers flow into the sound: the Housatonic, the Connecticut, and the Thames. These rivers are estuarine in their lower reaches and there are narrow marshes bordering them. Gardiners Bay on the northeast end of Long Island is very shallow, is well-protected from open-ocean influence, and is bordered by marine marsh. Block Island Sound, east of Long Island Sound, is essentially unprotected and is directly affected by the open Atlantic as well as receiving abundant water from Long Island Sound.

Parker's "Facies 1" is in the lower part of the Connecticut and Housatonic rivers which empty into Long Island Sound, and consists entirely of arenaceous Foraminifera and Thecamoebina characteristic of marshes, estuaries, and other low-salinity environments. Species characteristic of "Facies 1" also occur in Long Island Sound near river mouths. "Facies 2" occurs in Long Island Sound, Gardiners Bay, and Buzzards Bay, all areas having some protection from influences of undiluted water from the open Atlantic. "Facies 3" is characteristic of those areas which are exposed to open-ocean influences, Block Island Sound and south of Cuttyhunk. Both Facies 2 and 3 are dominated by calcareous forms. The faunas are listed in Figures 49 and 50.

COASTAL LAGOONS OF TEXAS

A comprehensive study has been made of the distribution of Foraminifera in an area centering around San Antonio Bay, Texas (Parker *et al.*, 1953). The geography and distribution of foraminiferal assemblages in this area are shown on Figure 51. The bays are very shallow and are separated from the open Gulf of Mexico by almost continuous barrier islands composed of sand. Narrow inlets provide avenues for interchange of water between the bays and the Gulf. The major faunal pattern in this area is a bay (lagoon) fauna and an open-gulf fauna. On the delta of the Guadalupe River both a

	FACIES 1		FACIES 2			FACIES 3A	FACIES 3B	
PERSISTENT OCCURRENCE ——— SCATTERED OCCURRENCE - - - SINGLE OCCURRENCE -	HOUSATONIC RIVER	CONNECTICUT RIVER	LONG ISLAND SOUND	GARDINERS BAY	BUZZARDS BAY	TRANSITION ZONES	BLOCK ISLAND SOUND	S.W. OF CUTTYHUNK
Ammoastuta salsa		- - -	-			-		
Ammobaculites cf. dilitatus						- - -		
A. cf. exiguus	-	- - -				-		
A. cf. foliaceus							- - -	
Ammoscalaria fluvialis	-	- - -				-		
Bolivina pseudoplicata			- - -		- - -			
B. variabilis		-	- - -	- - -	- - -		———	
Bulimina aff. aculeata							- - -	- - -
Cibicides concentricus							- - -	
C. lobatulus			- - -		-		———	- - -
Discorbis columbiensis			- - -	- - -	- - -			
D. squamata							- - -	-
Eggerella advena		- - -	———					
Elphidium advenum	-		-	- - -			-	
E. advenum var. margaritaceum		-	- - -	- - -				
E. excavatum			- - -	- - -			- - -	-
E. incertum		- - -						
E. incertum (heavy shell)							- - -	- - -
E. selseyense						- - -		
E. subarcticum	- - -							
Eponides frigidus							- - -	
E. frigidus var. calidus	- - -	-						- - -
E. wrightii			-	- - -	- - -		———	-
Globulina caribaea			- - -				- - -	-
Glomospira gordialis							- - -	-
Hopkinsina pacifica atlantica			- - -	- - -	- - -			
Labrospira crassimargo								———
Lagunculina vadescens		- - -						
Leptodermella variabilis	-	———						
Miliammina fusca	- - -		-		- - -	-		

Figure 49. Generalized distributions of benthonic Foraminifera in Long Island Sound—Buzzards Bay area. After F. L. Parker (1952).

	FACIES 1		FACIES 2				FACIES 3 A	FACIES 3 B
——— PERSISTENT OCCURRENCE - - - SCATTERED OCCURRENCE - SINGLE OCCURRENCE	HOUSATONIC RIVER	CONNECTICUT RIVER	LONG ISLAND SOUND	GARDINERS BAY	BUZZARDS BAY	TRANSITION ZONES	BLOCK ISLAND SOUND	S.W. OF CUTTYHUNK
Nonion tisburyense		- - -				- - -	- - -	
Nonionella atlantica						- - -	-	
Pninaella pulchella						- - -		
Poroeponides lateralis			- - -		-		———	- - -
Proteonina atlantica			- - -			———		
P. hancocki	-	———						
P. lagenarium		———	- - -					
P. sp. A			- - -			———	-	
P. sp. B		———	- - -					
Pseudopolymorphina novangliae			- - -	- - -	- - -	———	- - -	- - -
Pyrgo striatella						———	———	- - -
Quinqueloculina seminula			- - -	- - -	- - -	———	———	- - -
Q. seminula var. jugosa			- - -	-	- - -	———	———	- - -
Q. subrotunda			- - -	-		———	———	- - -
Reophax curtus						- - -		
R. dentaliniformis			———		———	- - -		
R. nana			- - -				- - -	
Rotalia beccarii			- - -					
R. beccarii var. sobrina			———		———		- - -	
R. beccarii var. tepida			- - -	- - -	- - -	———	-	
Textularia cf. tenuissima		-	- - -	- - -		———	-	- - -
Triloculina brevidentata				- - -	———			
Trochammina compacta			- - -	- - -	———	———	- - -	- - -
T. inflata	- - -	-	-			- - -	- - -	
T. lobata		-	- - -	- - -	———			
T. macrescens	-	- - -	-			-	-	
T. squamata		———						
Urnulina compressa	-	———						
U. difflugaeformis		- - -						
Virgulina fusiformis						- - -	- - -	- - -

Figure 50. Generalized distributions of benthonic Foraminifera in Long Island Sound—Buzzards Bay area. After F. L. Parker (1952).

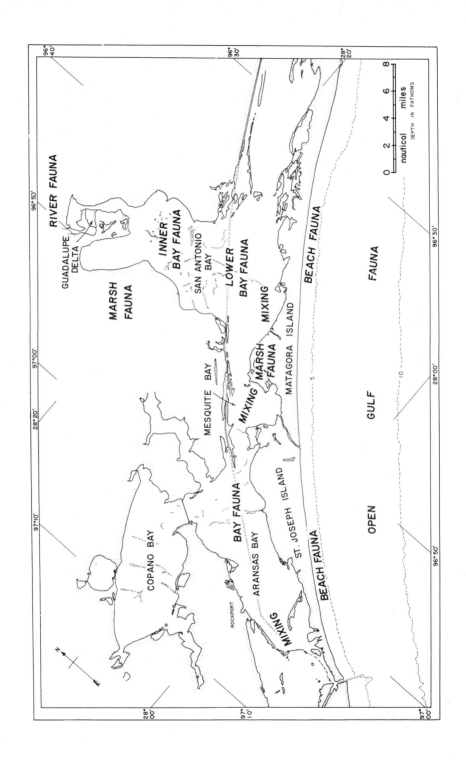

marsh fauna and a river fauna can be distinguished. The bay fauna can be divided into inner and lower bay assemblages. The marsh fauna also occurs in the small marshes on the landward side of the barrier island. The beach assemblage is closely related to, but distinguishable from, the open-gulf assemblage. The open-gulf assemblage occurs in all the lower bays of this area as a natural "contaminant" of the bay fauna; open-gulf species are most abundant near inlets, but also occur in some abundance throughout the lower bays. Only occasional specimens of bay species are recorded from the open-gulf fauna.

Laguna Madre, south of the San Antonio Bay area, is in large part hypersaline. It is very shallow and separated from the open ocean by a continuous sand barrier more than 100 mi. long. The foraminiferal fauna is very abundant and is a typical lower bay fauna, as in nearby Aransas Bay, but dominated by the Miliolidae, which constitute 40-90% of the population (Phleger, 1960b).

Faunas of Matagorda Bay have been reported by Shenton (1957) and Lehmann (1957). Lehmann has differentiated several foraminiferal "facies" by statistical techniques, as follows: Littoral beach, Offshore-bar marsh, "Back" bay type 1, "Back" bay type 2, "Fore" bay type 1, "Fore" bay type 2, Bay delta edge type 1, Bay delta edge type 2, Delta mainland marsh pond, and Marginal brackish lake. Lehmann's results emphasize the complexity of foraminiferal patterns in such a marginal marine area, and indicate that it should be possible to reconstruct the details of a Tertiary lagoon if sufficient samples are available. Statistical techniques such as these should be further explored. The major faunal pattern in Matagorda Bay is similar to that in the San Antonio Bay area.

Figure 51, Opposite. Distribution of foraminiferal faunas in San Antonio Bay area, Texas.

MISSISSIPPI SOUND AREA

This area includes all of Mississippi Sound between Mobile Bay and Gulfport, Mississippi, the adjacent marshes, and seaward to a depth of 10 fms. (18 m.) (Phleger, 1954a). The sound is separated from the open Gulf of Mexico by a series of barrier islands composed of sand. The islands are less continuous than those in the Texas area, and are separated by a series of relatively wide inlets. The geography and distribution of faunas in the Mississippi Sound area are shown on Figure 52.

Mississippi Sound contains an endemic fauna which appears to be composed mostly of arenaceous Foraminifera, principally *Ammotium*. The open-gulf fauna seaward from the barrier islands contains mostly calcareous species. The boundary between the sound and open-gulf faunas is relatively sharp at the barrier islands. The open-gulf fauna has invaded the sound for a short distance inside the inlets and has spread laterally behind the barrier islands; one invasion of the open-gulf fauna extends to the mainland. Where the open-gulf species occur within the sound they are in rather discrete areas and are mixed with the *Ammotium* fauna. Few specimens from the sound fauna are recorded seaward from the barrier islands except very nearshore and mostly near inlets. The beach and very nearshore fauna on the barrier islands generally is distinctive. The marsh bordering Mississippi Sound on the mainland has a characteristic marsh fauna, and the estuary fauna can be distinguished from it. Estuary and marsh species contaminate the sound fauna on the mainland side of the sound and also near well-developed areas of marsh on the islands.

CHAPTER III 135

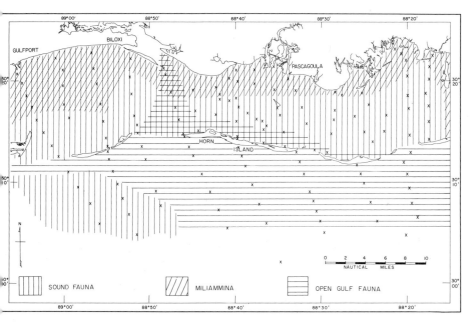

Figure 52. Faunas of benthonic Foraminifera in the Mississippi Sound area, Mississippi. After Phleger (1954a).

MISSISSIPPI DELTA AREA

A series of faunas similar to those in the Mississippi Sound area have been described from the southeastern Mississippi Delta (Phleger, 1955a; Lankford, 1959), shown on Figure 53. In this area Breton Sound is only partially separated from the open Gulf of Mexico by a small, low barrier island. A very wide inlet between the island and the delta permits relatively unrestricted flow of water. This is an area of high runoff, of sedimentary aggradation, and of extensive marsh.

The pure open-gulf fauna is restricted to the area seaward from the barrier island and there is a transition area between open-gulf and sound faunas several miles in width. The sound fauna is similar to that in the San Antonio Bay area except for the presence of abundant marsh specimens, especially near the shore.

Lankford (1959) recognizes the following additional faunas from the Mississippi Delta margin: interdistributary bay in the bays between the river passes, fluvial marine in the lower part of the river, and deltaic marine on the pro-delta margin where sedimentation is fast and there is active mixing of river and marine water. Warren (1956) has described faunas of polyhaline lakes and bays, marsh, and nearshore gulf in the Mississippi Delta area. Treadwell (1955) has recognized marsh, inner lake, outer lake, sound, and beach faunas in the area adjoining Chandeleur Sound.

Interpretation of Marginal Marine Faunal Patterns

The nearshore distributions of benthonic Foraminifera which have been described from these areas obviously are conditioned by many or most of the ecologic factors discussed in the previous chapter. These distributions are not depth assemblages in the sense of the depth distributions discussed, with possible exceptions. The basic features which require explana-

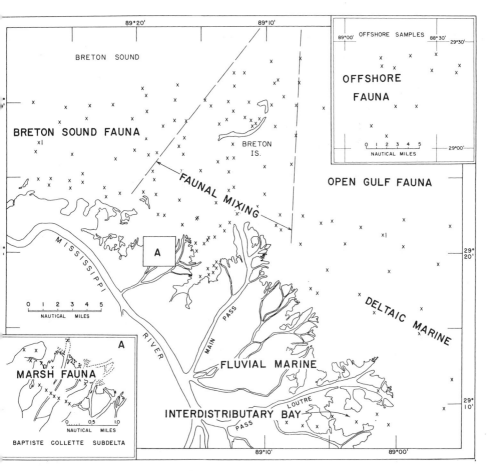

Figure 53. Faunas of benthonic Foraminifera in southeastern Mississippi Delta area, Louisiana. After Phleger (1955a) and Lankford (1959).

tion are the differences between the lagoon and the open-ocean faunas, and the differences and similarities between lagoon faunas in different areas.

These distributions can be better understood in terms of oceanography if we adopt a general ecologic concept instead of attempting to sort out specific ecologic factors. The adaptation of different assemblages of organisms to different types of marine water was mentioned in Chapter I, and this concept is useful in interpreting the nearshore distributions which have been described.

It is suggested that the difference between the lagoon and open-ocean faunas is basically due to barriers which prevent invasion of unaltered open-ocean water. These are both physiographic barriers, such as islands and bars, and dynamic water barriers caused by land runoff. The effect is to impede constant flow of typical oceanic water into the lagoon. In areas of no runoff, such as Laguna Madre, the properties of the water are altered by evaporation. It seems logical that we need to consider principally the bottom water since we are dealing with benthonic forms although the entire water column must have some effect.

MISSISSIPPI SOUND AREA

In the Mississippi Sound area (Figure 52) the sound or lagoon fauna consists mostly of arenaceous forms and most of the open-ocean calcareous species which occur there appear to be associated with the inlets as discussed above. The islands act as a barrier to the invasion of open-gulf water along a broad front and the only possible route is through the inlets. The high runoff from the adjacent mainland, however, forms an excess of lagoon water which is a dynamic barrier. This lagoon water is of intermediate salinity, 20-30 o/oo, and is formed by mixture of the fresh river runoff with the open-ocean water invading through the inlets. The mixing probably occurs near the inlets and is aided by wind causing water turbulence at the shallow depths which prevail in Mississippi Sound. Since there is an excess of Mississippi Sound water it flows out above the more saline and denser open-ocean water and eventually is mixed with open-ocean water to become a part of the inner continental shelf water (see Figure 54a).

The nearshore open-gulf water can be recognized immediately above the bottom just outside the barrier islands and in occasional bottom-water samples taken just inside the inlets. This water has a salinity of 33-35 o/oo, on the average, and underrides the Mississippi Sound water because of its greater density, invading Mississippi Sound along the bottom. Where this water is present in the sound over a period of time or at frequent intervals it provides an environment suitable for the open-ocean benthonic Foraminifera which migrate into the sound along with it. These open-ocean benthonic forms probably invade Mississippi Sound because of the environment created by this bottom water, and not principally because of physical transport. At least some of these species appear to be able to tolerate considerable variation in their environment, on the basis of what is known of their distribution and physiology. Examples of such forms are *Streblus beccarii* (Linné) variants and several species of *Elphidium*. Their distribution in the Mississippi Sound area suggests that the principal adjustment may be to the water occurring on the inner continental shelf in this area and not to the Mississippi Sound water.

Distributions of open-ocean benthonic Foraminifera in Mississippi Sound reflect the routes of invasion of open-ocean water along the bottom. Such distributions may be a result of the average conditions. A direct survey of water characteristics at any particular time, on the other hand, measures only those conditions obtaining when the survey was made and

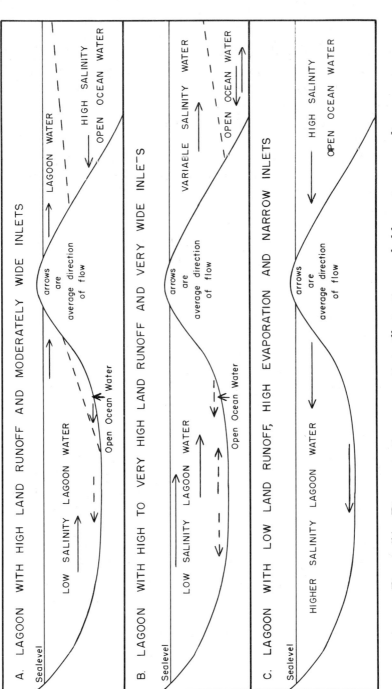

Figure 54. Diagrammatic cross-sections illustrating probable water mass distributions and directions of flow in coastal lagoons having different amounts of land runoff.

cannot measure average conditions. Distributions of these open-ocean Foraminifera in Mississippi Sound indicate invasion of unmixed gulf water along the bottom for approximately four miles inside the inlets. The samples extending north of the inlet at the western end of Horn Island (Figure 52) suggest at least occasional invasions almost to the shore of the mainland. The fact that unmixed gulf water does not invade the inlet at the eastern end of Horn Island for a greater distance may be due to the outflow of runoff from the Pascagoula River in this area. The presence of open-gulf species behind the islands suggests that the nearshore, open-ocean water spreads laterally into these areas and becomes mixed along a broad and irregular front.

Where open-ocean species occur in Mississippi Sound they are mixed with the *Ammotium* fauna characteristic of the sound. It is of interest that there is essentially no contamination of the open-ocean fauna with species from Mississippi Sound at stations outside the barrier islands. This is believed to reflect the vertical distribution of the lighter-density water from Mississippi Sound overriding the denser open-ocean water outside the barriers (see Figure 54a). Under such hydrographic conditions the typical sound Foraminifera generally do not escape because their adjustment is to a type of water affecting the bottom where they live. In the southwesternmost part of the open-ocean area, however, there are several stations containing a high frequency of the *Ammotium* fauna (see Figure 52). This suggests rather persistent presence of relatively unmodified sound water throughout the entire water column at this locality.

The occurrence of *Miliammina* on the inner part of Mississippi Sound reflects the presence of continuous marine marshes which fringe the mainland and are especially developed near large streams.

TEXAS AREA

The pattern of open-ocean and lagoon distributions in the San Antonio Bay area (Figure 51) is similar to that in Mississippi Sound and most of the species are common to the two areas. There are some differences in composition, however, between the lagoon fauna of San Antonio Bay and that of Mississippi Sound. In the San Antonio Bay area the lagoon fauna contains *Ammotium*, but it is not the dominant element of the population except near the Guadalupe River. The typical open-ocean assemblage also is present in this lagoon, but it is more widespread than in Mississippi Sound. *Streblus beccarii* (Linné) variants and *Elphidium gunteri* Cole are the two dominant species in all the bays in the Texas area except in parts of Laguna Madre. These two species are less abundant in Mississippi Sound.

These differences in distribution are surprising since few, relatively narrow inlets breach the barrier islands in the San Antonio Bay area. A possible explanation is that most of the water in the San Antonio Bay area lagoons is supplied from the open ocean much of the time. This is an area of low and erratic rainfall and runoff. During long dry periods most of the water in the bays appears to be oceanic water (salinity of 36 o/oo ±) entering through the passes. Excessive evaporation in the shallow, enclosed bays tends to produce salinities of higher values than open-gulf water. High salinities (up to 42 o/oo) obtained during July and August, 1951, according to F. L. Parker *et al.* (1953) are due to this mechanism (see Figure 54c). In Laguna Madre the water is always hypersaline except near the inlets, and the lagoon *Ammotium* is rare.

It is believed that this general and widespread invasion of open-ocean water during the long dry periods explains the lagoon distributions of the species which are essentially open-ocean forms in the Mississippi Sound area. Open-ocean species are expected to migrate along with the water mass to which they are ecologically adjusted. Abundant forms such as *Elphidium gunteri* Cole and *Streblus beccarii* (Linné) variants appear to be able to withstand times of high runoff and lowered lagoon salinities (see Figure 43); these forms proliferate, probably because of little competition from other species. It is possible that much or most of their reproduction occurs when the open-ocean water is in the lagoon. The "endemic" fauna of *Ammotium* may be able to reproduce principally during times of high runoff and lowered salinities. Its absence from the open-ocean stations suggests that this population does not survive in the open-ocean water for an indefinite period.

It is believed that distributions in nearby Matagorda Bay, and associated bays, may be caused by factors similar to those in San Antonio Bay. Faunas described by Lehmann (1957) and Shenton (1957) are similar in composition to lower San Antonio Bay faunas.

MISSISSIPPI DELTA AREA

The faunal boundaries are much less sharp in the southeastern Mississippi Delta area than in Mississippi Sound and the Texas bays. The average boundary between the open-ocean and lagoon faunas is a transitional one over a distance of approximately 8-10 miles. The more abundant forms in the open-ocean fauna in the Mississippi Sound area, such as *Elphidium gunteri* Cole, *E.* spp., and *Streblus beccarii* (Linné) variants, also occur in abundance in the lagoon faunas. Likewise, the lagoon *Ammotium*, particularly *A. salsum* (Cushman and Brönnimann), is rather widespread in the nearshore open-gulf fauna.

These distributions seem to reflect the erratic and irregular distributions of the Breton Sound and open Gulf of Mexico water in the area. The open-gulf water is more saline and dense than water from either the sound or the Mississippi and tends to invade Breton Sound along the bottom, as demonstrated by vertical salinity distributions obtained by Scruton (1956). The Breton Sound water of intermediate salinity (25-29 o/oo) is formed by mixing with river water, primarily from Main Pass in the area studied. The extent of relatively unmodified open-gulf water invading along the bottom is expected to show great variation, depending upon wind force and direction, tides, and runoff from the river. During onshore wind and low runoff from Main Pass, relatively undiluted gulf water may extend for a considerable distance into Breton Sound. During very high discharge from Main Pass, on the other hand, there may be little or no undiluted open-gulf water in or near the southern end of Breton Sound. The occurrence of *Ammotium salsum* (Cushman and Brönnimann) in some abundance for several miles outside the inlets suggests that water of intermediate salinity occasionally extends from the surface to the bottom in this open-gulf area.

It is surprising that open-ocean and lagoon faunas of benthonic Foraminifera can be distinguished in an area such as this where the physical barrier of Breton Island is so small compared to the width of the inlets. The abundant runoff from the river distributaries provides an excess of sound water of intermediate salinity especially during times of high runoff. This acts as a dynamic water barrier which tends to prevent unrestricted flow of open-ocean water into Breton Sound, aided by the presence of Breton Island and associated sandbars (see Figure 54b).

The fauna in Breton Sound is similar to the lagoon, or bay, fauna in the San Antonio Bay area. The reasons for the presence of a similar fauna are not the same in the two areas, however, and reflect different distributions of barrier islands as well as amounts of land runoff. The lagoon fauna in the Delta area differs from the San Antonio Bay fauna in one important respect: marsh Foraminifera are quite widespread in Breton Sound as apparent contaminants of the sound fauna. This reflects the widespread marine marshes characteristic of an active delta such as that of the Mississippi River.

OTHER DISTRIBUTIONS

The salinities in Long Island Sound recorded by F. L. Parker (1952, p. 432) show a progressive increase from central Long Island Sound at a minimum of 28 o/oo to a maximum of about 33 o/oo in Block Island Sound. There is a decrease to a minimum of about 30 o/oo in upper Buzzards Bay. Both Long Island Sound and Buzzards Bay have large rivers entering them and relatively low-salinity water is formed in abundance. This distinctive water appears to provide the environment for Parker's "Facies 2." Her "Facies 3" is developed where the influence of the open Atlantic upper shelf water predominates (see Figure 48).

The differentiation of the lagoon and open-ocean faunas in the Barnstable Harbor area is more difficult to explain on the basis of different bodies of water. Most of the water which enters this lagoon area is brought in by tidal currents, and there is little surface runoff on a continuous basis. There is probably some addition of fresh water from ground water, and also during rather frequent rains. The recorded salinities on the intertidal flats are 25-31 o/oo, and the salinity of the open ocean in this area is rather constant at 31-32 o/oo.

Relationship between Faunal Patterns, Relative Runoff and Extent of Barrier Islands in the Northern Gulf of Mexico

The generalizations presented above for Mississippi Sound, San Antonio Bay, Laguna Madre, Matagorda Bay, and the Mississippi Delta may be applicable to other areas. Diagrammatic cross-sections (Figure 54) show the generalized directions of water flow for an area of high runoff, one of very high runoff and one of low runoff. These sections are made through the barrier islands and the inlets are not shown; most or all of the flow is through the inlets, and it is not intended to imply in these diagrams that the flow is through the sand barrier islands.

These different faunal patterns in the northern Gulf of Mexico appear to reflect differences in the amount of runoff in the various areas. It is also of interest to observe that the size and continuity of the sand barrier islands seem to have an inverse relationship to the relative amount of runoff. The area with the greatest runoff, the southeastern Mississippi Delta (Figure 53), has the smallest island barrier. In the San Antonio Bay area (Figure 51) and Laguna Madre, on the other hand, with the least runoff, the barrier islands are essentially continuous with inlets so narrow that some are opened only by severe storms, floods, or dredging.

The amount of water which flows in and out of a lagoon or bay through an inlet or entrance due to rise and fall of tides has been called the "tidal prism." Tidal prism is measured by the difference between high and low tide within a lagoon multiplied by the area, and volume has been expressed by coastal engineers in acre feet of water.

It has been shown that on the Pacific coast of the United States there is a relationship between the tidal prism and the cross-section of entrance channels to harbors which will maintain themselves (O'Brien, 1931). Robins (1933) has indicated that one acre foot of tidal prism will maintain about one square foot of channel cross-section. On the United States Gulf Coast lunar tides are of small magnitude but wind-driven tides occasionally cause water level changes of a few feet. Runoff in some places in the Gulf Coast has the same effect on inlets as tidal flushing in areas of higher tidal range.

It appears that the width of the inlets along the Gulf Coast is more or less related to the excess of intermediate lagoon water which is formed by mixing of runoff and ocean water. The abundant flow of such water seaward probably keeps the inlets open and counteracts the action of storm waves and consequent longshore currents which tend to close them. Where there is a great excess of runoff, such as off Main Pass in the Mississippi Delta area, the size of the barrier islands tends to be small (Figure 53). Breton Island, the barrier in this area, shows this situation where the island-forming material has little opportunity to become stabilized. The almost continuous Chandeleur Islands lying to the north of Breton Island may be a result of insufficient excess of Chandeleur Sound water to open significant inlets across the islands. This is an old delta area now inactive, and it is to be presumed that Breton Island might become considerably extended if the active Mississippi Delta were to move southward and/or westward.

The inlets between the islands on the seaward side of Mississippi Sound (Figure 52) are not so wide as those bordering Breton Island. While the runoff into Mississippi Sound is considerable, it is much less than that into Breton Sound. The inlet at the north end of Chandeleur Sound may be caused by a large outflow of intermediate water at this location. The occurrence of the lagoon *Ammotium* fauna in the open-ocean sediments several miles seaward from this inlet indicates the outflow of a considerable amount of lagoon-type water in this area.

This line of reasoning suggests that the total outflow of lagoon water from the San Antonio Bay area for long periods of time is so insignificant that few inlets can be maintained (Figure 51). During infrequent times of flood or of severe storms some passes may be opened or enlarged; these may tend to heal themselves because of lack of continuous vigorous flushing action to keep them free of sediment. The few, narrow passes which do exist probably are just sufficient to admit inflow of oceanic water and permit tidal flow. This is similar to the situation at Barnstable Harbor, Massachusetts (Figure 47) where there is also no excess of lagoon water formed (Phleger and Walton, 1950), and although there is a tidal range of several feet the tidal prism is small because the area of the lagoon is small. Sandy Neck, a spit, has almost closed the inlet to Barnstable Harbor.

Laguna Madre has a long, continuous sand barrier, Padre Island. This is an area where there is essentially no runoff most of the time, and inlets are not required to accommodate water outflow. The only breach in Padre Island is at the position of the Rio Grande.

Composition of Gulf Coast and Eastern North American Nearshore Faunas

Figures 55-56 show the generalized distributions of the useful species which have been reported from most of the nearshore areas discussed above. In addition, lists of species in the various faunas are given below. Faunas from other areas are listed later in this chapter. It seems probable that faunas similar to these occur elsewhere in similar environments, but the specific compositions may be somewhat different. Not all species recorded are listed, and the species listed do not occur in all the areas described.

Figure 55, Opposite. Generalized distributions of nearshore benthonic Foraminifera. Heaviness of the lines indicates relative abundance.

Species	OPEN OCEAN					LAGOON					ESTUARY			MARSH		
	BARN-STABLE HARBOR AREA	LONG IS. SOUND AREA	SAN ANTONIO BAY AREA	MISS. SOUND AREA	MISS. DELTA AREA	BARN-STABLE HARBOR AREA	LONG IS. SOUND AREA	SAN ANTONIO BAY AREA	MISS. SOUND AREA	MISS. DELTA AREA	LONG IS. SOUND AREA	SAN ANTONIO BAY AREA	MISS. SOUND AREA	BARN-STABLE HARBOR AREA	SAN ANTONIO BAY AREA	MISS. SOUND AREA	MISS. DELTA AREA
Ammobaculites cf. foliaceus (H.B. Brady)	–																
Ammodiscus spp.						•••••											
Bigenerina irregularis Phleger & Parker					–			–	•••••								
Bolivina lowmani Phleger & Parker									•••••								
B. pseudoplicata Heron-Allen & Earland			–				–										
B. pulchella primitiva Cushman				•••••													
B. striatula Cushman					–												
B. variabilis (Williamson)				•••••							•••••						
Buccella hannai (Phleger & Parker)				•••••	–												
Bulimina aff. aculeata d'Orbigny					–				•••••	•••••							
Buliminella cf. bassendorfensis Cushman & Parker					–			–	•••••	•••••							
B. elegantissima (d'Orbigny)					–			–	•••••	•••••							
Cibicides lobatulus (Walker & Jacob)			•••••	–	–			–	•••••	–					•••••		
Cibicidina strattoni (Applin)								–	•••••								
Discorbis cf. columbiensis Cushman				–				–		–					•••••		
D. floridana Cushman																	
D. squamata Parker																	
Elphidium articulatum d'Orbigny		–															
E. discoidale (d'Orbigny)			•••••					–	•••••	–							
E. incertum (Williamson) & variants													─	•••••	–	–	
E. spp.									•••••								•••••
Eponides wrightii (H.B. Brady)				•••••	–		–										
Gaudryina exilis Cushman & Bronnimann			–		–			–		–							
Globulina caribaea d'Orbigny							–										
Glomospira cf. gordialis (Jones & Parker)		–		•••••													
Guttulina australis (d'Orbigny)			•••••	–					•••••								
Labrospira crassimargo (Norman)																	
Miliolidae						•••••											
Nonionella atlantica Cushman		–	–		–			–	•••••						•••••		
N. opima Cushman					–			–									
Nouria polymorphinoides Heron-Allen & Earland			–														
Poroeponides lateralis (Terquem)			–	–													
Proteonina atlantica Cushman				•••••													
Pyrgo striatella (Cushman)								•••••									
Quinqueloculina cf. compta Cushman			–					–		–							
Q. cultrata (H.B. Brady)			•••••		–			–									
Q. funafutiensis (Chapman)			–					•••••									
Q. lamarckiana d'Orbigny			–					•••••									
Q. poeyana d'Orbigny								•••••									
Q. seminula (Linné) & variants								•••••							–	–	
Q. subrotunda (Montagu)								•••••									
Reophax curtus Cushman																	
R. scottii Chaster																	
R. scorpiurus Montfort	•••••																
Reussella atlantica Cushman			–	•••••													
"Rotalia" pauciloculata Phleger & Parker			–	•••••													
"R." rolshauseni Cushman & Bermudez				•••••	•••••												
Spiroplectammina biformis (Parker & Jones)	–	–															
Textularia cf. mayori Cushman			•••••	•••••													
Triloculina trigonula (Lamarck)								•••••									
Virgulina pontoni Cushman								•••••									

Species	Open Ocean Barnstable Harbor	Open Ocean Long Is. Sound	Open Ocean San Antonio Bay	Open Ocean Miss. Sound	Open Ocean Miss. Delta	Lagoon Barnstable Harbor	Lagoon Long Is. Sound	Lagoon San Antonio Bay	Lagoon Miss. Sound	Lagoon Miss. Delta	Estuary Long Is. Sound	Estuary San Antonio Bay	Estuary Miss. Sound	Marsh Barnstable Harbor	Marsh San Antonio Bay	Marsh Miss. Sound	Marsh Miss. Delta
Buccella frigida (Cushman)	—																
Eggerella advena (Cushman)		—												•••••			
Elphidium advena (Cushman)		—					—				—			•••••			
E. advenum margaritaceum Cushman							—										
E. gunteri Cole												•••••					
E. cf. koeboeense LeRoy			—												—		
E. poeyanum (d'Orbigny)			—												—		
E. subarcticum Cushman			—														
Massilina peruviana (d'Orbigny)																	
"Rotalia" beccarii (Linné) variants			—					•••••			•••••					•••••	•••••
Trochammina lobata Cushman														•••••			
T. squamata Parker & Jones							—				—			•••••			
Ammobaculites dilatatus Cushman & Bronn.			—							•••••							
A. exilis Cushman & Bronnimann			•••••		•••••					—							
A. salsus Cushman & Bronnimann variants			•••••		•••••					—		•••••		—	—		
A. spp.								—		—					—		
Elphidium delicatulum Bermudez				•••••						—					—		
E. matagordanum (Kornfeld)			—							—					—		
Nonion tisburyense Butcher		—	•••••								—						
Quinqueloculina rhodiensis Parker																	
Reophax dentaliniformis H.B. Brady																	
R. nana Rhumbler		—	—														
"Rotalia" beccarii (Linné) variant C											•••••						
Textularia tenuissima Earland																•••••	
Triloculina brevidentata Cushman																	
T. sidebottomi (Martinotti)																	
Triloculinella obliquinoda Riccio		—									—						
Trochammina compacta Parker											—						
Ammoastuta inepta (Cushman & McCulloch)		•••••					•••••										
Ammobaculites cf. exiguus Cushman & Bronnimann		•••••															
Ammoscalaria fluviatilis Parker		•••••															
Arenoparrella mexicana (Kornfeld)																	
Armorella sphaerica Heron-Allen & Earland																	
Discorinopsis aguayoi (Bermudez)																	
Haplophragmoides subinvolutum Cush. & McCull.																	
Lagunculina vadescens Cushman & Bronnimann																	
Leptodermella variabilis Parker																	
Miliammina fusca (H.B. Brady)	—		•••••														
Palmerinella palmerae Bermudez																	
Proteonina hancocki Cushman & McCulloch																	•••••
P. lagenaria (Berthelin)																	
Recurvoides sp.													•••••				
Trochammina comprimata Cushman & Bronn.													•••••				
T. inflata (Montagu)	•••••	—			•••••								•••••			•••••	
T. macrescens H.B. Brady		—														•••••	
Uruulina compressa Cushman																	—
U. diffulgiformis Gruber															•••••		
U. spp.								•••••					•••••				

OPEN-OCEAN FAUNA

The following forms seem to be relatively important in the open-ocean nearshore fauna (less than 20 m.) at many localities (see Plate 7 for some of these species):

Alveolophragmium crassimargo (Norman)
Ammoscalaria cf. *A. foliaceus* (H. B. Brady)
Bigenerina irregularis Phleger and Parker
Bolivina lowmani Phleger and Parker
B. pseudoplicata Heron-Allen and Earland
B. pulchella primitiva Cushman
B. striatula Cushman
B. variabilis (Williamson)
Buccella hannai (Phleger and Parker)
Buliminella cf. *B. bassendorfensis* Cushman and Parker
B. elegantissima (d'Orbigny)
Cibicides lobatulus (Walker and Jacob)
Elphidium discoidale (d'Orbigny)
E. incertum (Williamson) and variants
Gaudryina exilis Cushman and Brönnimann
Glabratella wrightii (Brady)
Glomospira cf. *G. gordialis* (Jones and Parker)
Guttulina australis (d'Orbigny)
Hanzawaia strattoni (Applin)
Miliolidae (various spp.)
Nonionella atlantica Cushman
N. opima Cushman
Poroeponides lateralis (Terquem)
Proteonina atlantica Cushman
Reophax curtus Cushman
R. scottii Chaster

Figure 56, Opposite. Generalized distributions of nearshore benthonic Foraminifera. Heaviness of the lines indicates relative abundance.

Reussella atlantica Cushman
Rosalina cf. *R. columbiensis* (Cushman)
R. floridana (Cushman)
R. squamata (Parker)
"*Rotalia*" *pauciloculata* Phleger and Parker
"*R.*" *rolshauseni* Cushman and Bermudez
Textularia cf. *T. mayori* Cushman
Virgulina pontoni Cushman

Many of these species also occur in the lagoon fauna in certain areas, but in lower frequencies than in the open ocean. These occurrences in the lagoon assemblage may be thought of as "natural contamination" or gradation between faunas.

LAGOON FAUNA

The following species appear to be characteristic of lagoon assemblages in the areas reported; many of these species are illustrated on Plate 8. Most of the lagoon species seem to occur also in the marsh and estuary deposits, but in lower frequencies. They also are reported from many open-ocean stations in very low frequencies and are interpreted as "contamination" at such stations.

Ammobaculites dilatatus Cushman and Brönnimann
Ammotium salsum (Cushman and Brönnimann) variants
Elphidium delicatulum Bermúdez
E. matagordanum (Kornfeld)
Nonion tisburyensis Butcher
Reophax dentaliniformis Brady
R. nanus Rhumbler
Streblus beccarii (Linné) variant C
Textularia earlandi Parker
Trochammina compacta Parker

ESTUARY AND SALT-MARSH FAUNAS

The following species are characteristic of faunas in small estuaries and marshes; illustrations of some of the species are on Plate 8. These two assemblages are grouped together here because it is often rather difficult to separate them faunally. It is apparent that there are several species which may be endemic to the marsh but are commonly washed into the estuarine environment because of the intimate association of these environments. Such species are found in much lower frequencies in the estuarine fauna than in the adjacent marsh. It is difficult in the field to place a boundary between these two environments because they intergrade; it may not be practical to attempt to separate them in paleoceanographic studies. Marsh and estuary species occur in low frequencies in many lagoon stations near estuaries and marshes.

Ammoastuta inepta (Cushman and McCulloch)
Ammoscalaria fluvialis Parker
Arenoparrella mexicana (Kornfeld)
Discorinopsis aguayoi (Bermúdez)
Haplophragmoides subinvolutum Cushman and McCulloch
Miliammina fusca (Brady)
Palmerinella palmerae Bermúdez
Recurvoides sp.
Tiphotrocha comprimata (Cushman and Brönnimann)
Trochammina inflata (Montagu)
T. macrescens Brady

SPECIES COMMON TO MORE THAN ONE ENVIRONMENT

The following species occur in both the open-ocean and lagoon assemblages in some places, and some are occasionally

(Text continued on page 158).

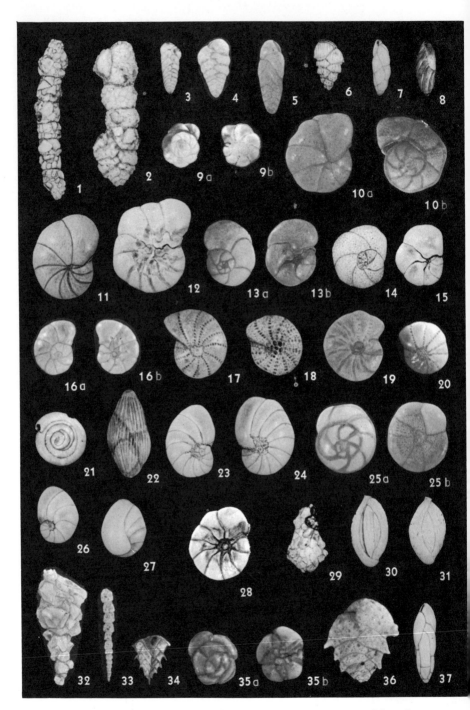

Plate 7. Some common, nearshore open-ocean species of benthonic Foraminifera.

CHAPTER III 155

EXPLANATION OF PLATE 7

1, 2. *Bigenerina irregularis* Phleger and Parker. × 29.
3. *Bolivina lowmani* Phleger and Parker. × 53.
4. *Bolivina pseudoplicata* Heron-Allen and Earland. × 53.
5. *Bolivina variabilis* (Williamson). × 53.
6. *Bolivina pulchella primitiva* Cushman. × 53.
7. *Buliminella* cf. *B. bassendorfensis* Cushman and Parker. × 47.
8. *Buliminella elegantissima* (d'Orbigny). × 56.
9a, b. *Buccella hannai* (Phleger and Parker). × 36.
10a, b. *Cibicides lobatulus* (Walker and Jacob). × 38.
11, 12. *Hanzawaia strattoni* (Applin). (11) × 32. (12) × 29.
13a, b. *Rosalina columbiensis* (Cushman). × 38.
14, 15. *Rosalina floridana* (Cushman). × 47.
16a, b. *"Discorbis" squamata* Parker. × 53.
17. *Elphidium discoidale* (d'Orbigny). × 32.
18. *Elphidium gunteri* Cole. × 32.
19. *Elphidium incertum* (Williamson) variant. × 38.
20. *Elphidium incertum mexicanum* Kornfeld. × 36.
21. *Glomospira* cf. *G. gordialis* (Jones and Parker). × 32.
22. *Guttulina australis* (d'Orbigny). × 32.
23, 24. *Nonionella atlantica* Cushman. × 47.
25a, b. *Glabratella wrightii* (Brady). × 38.
26, 27. *Nonionella opima* Cushman. × 47.
28. *Streblus beccarii* (Linné) var. A. × 38.
29. *Proteonina atlantica* Cushman. × 44.
30, 31. *Quinqueloculina compta* Cushman. × 32.
32. *Reophax curtus* Cushman variant. × 36.
33. *Reophax scottii* Chaster. × 53.
34. *Reussella atlantica* Cushman. × 36.
35a, b. *"Rotalia" rolshauseni* Cushman and Bermúdez. × 36.
36. *Textularia mayori* Cushman. × 36.
37. *Virgulina pontoni* Cushman. × 47.

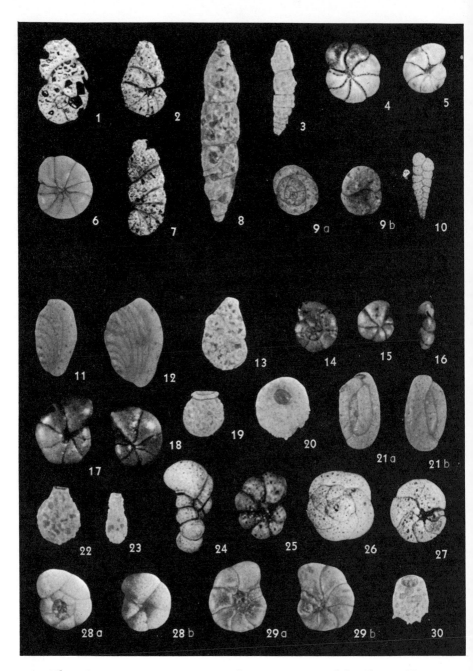

Plate 8. Figures 1-10, common lagoon species of benthonic Foraminifera; figures 11-30, common marsh and estuary species of benthonic Foraminifera and Thecamoebina.

CHAPTER III 157

EXPLANATION OF PLATE 8

   1. *Ammobaculites* cf. *A. dilatatus* Cushman and Brönnimann. × 36.
2 , 7. *Ammotium salsum* (Cushman and Brönnimann) vars. (2) ×56. (7) × 42.
   3. *Reophax nanus* Rhumbler. × 53.
   4. *Elphidium matagordanum* (Kornfeld). × 42.
   5. *Elphidium delicatulum* Bermúdez. × 56.
   6. *Nonion tisburyensis* Butcher. × 38.
   8. *Reophax dentaliniformis* Brady. × 38.
9a, b. *Trochammina compacta* Parker. × 53.
  10. *Textularia earlandi* Parker. × 53.
11 , 12. *Ammoastuta inepta* Cushman and McCulloch. × 53.
  13. *"Ammoscalaria fluvialis"* Parker. × 38.
14-16. *Arenoparrella mexicana* (Kornfeld). × 36.
17 , 18. *Haplophragmoides subinvolutum* Cushman and McCulloch. × 36.
  19. *Difflugia urceolata* Carter. × 53.
  20. *Centropyxis* (*Centropyxis*) sp. × 53.
21a, b. *Miliammina fusca* (Brady). × 53.
  22. *Pontigulasia compressa* (Carter). × 53.
  23. *Difflugia* sp. × 53.
24 , 25. *Recurvoides* sp. × 42.
26 , 27. *Tiphotrocha comprimata* (Cushman and Brönnimann). × 42.
28a, b. *Trochammina inflata* (Montagu). × 38.
29a, b. *Trochammina macrescens* Brady. × 38.
  30. *Centropyxis* (*Centropyxis*) sp. × 53.

present in the marsh-estuary fauna. It is suggested that most of these species may have their primary adaptation to the inner continental-shelf water, and are able to adapt themselves to variable conditions when this water occasionally invades other environments.

Buccella frigida (Cushman)
Eggerella advena (Cushman)
Elphidium advena (Cushman) and vars.
E. gunteri Cole
E. cf. *E. koeboeense* LeRoy
E. poeyanum (d'Orbigny)
E. subarcticum Cushman
Massilina peruviana (d'Orbigny)
Streblus beccarii (Linné) vars.
Trochammina lobata Cushman
T. squamata Parker and Jones

BEACH FAUNA

The beach fauna has been studied in detail in the San Antonio Bay area where it is especially characterized by an abundance of the following species which differentiate it from the open-ocean fauna (Plate 9):

Elphidium incertum mexicanum Kornfeld
Quinqueloculina cf. *Q. compta* Cushman
Q. seminulum (Linné)
Streblus beccarii (Linné) vars.

Quinqueloculina seminulum (Linné) is the most abundant form in this beach fauna. In the Mississippi Sound area many of the near-beach samples contain a small population composed mostly of *Quinqueloculina* and *Elphidium*. A beach fauna such as that present in the San Antonio Bay area may be due to physical sorting by wave action. *Quinqueloculina* is a relatively robust genus, and probably can withstand the violence of the surf zone better than most open-ocean species.

Numerous specimens from this environment appear to be eroded by surf-zone wear.

Boltovskoy (1955) has reported on the Foraminifera in beach sands at Quequén, Argentina, south of Buenos Aires. The most common species in this material are:

> *Quinqueloculina seminulum* (Linné)
> *Rotalia beccarii beccarii* (Linné) [=*Streblus*]
> *Buccella frigida* (Cushman)
> *Textularia gramen* d'Orbigny
> *Quinqueloculina intricata* Terquem
> *Elphidium discoidale* (d'Orbigny)

The following species also occur in appreciable frequencies:

> *Quinqueloculina lamarckiana* d'Orbigny
> *Q. angulata* (Williamson)
> *Triloculina subrotunda* (Montagu)
> *T. gibba* d'Orbigny
> *Pyrgo patagonica* d'Orbigny
> *P. nasuta* Cushman
> *Spiroloculina planulata* Lamarck

This fauna is dominated by large Miliolidae and *Textularia*.

It has been pointed out previously that characteristics of beach and nearshore sediment are the relatively coarse size and good sorting caused by the intensive turbulence. Under such conditions it would be expected that only those Foraminifera larger than a certain size could remain with the sediment, and the smaller tests would be transported out of the beach and nearshore zone and deposited elsewhere. Smaller, fragile forms also may be destroyed under such conditions.

FLUVIAL MARINE FAUNA

This association is reported by Lankford (1959) from the lower distributaries of the Mississippi River where there is an invasion of a salt-water wedge along the bottom. *Palmerinella gardenislandensis* (Akers) is the characteristic species. At or near the mouths of the distributaries there are also *Bolivina lowmani* Phleger and Parker, *Nonionella opima* Cushman, *Epistominella vitrea* Parker, *Buliminella* cf. *B. bassendorfensis* Cushman and Parker, and a few other rare species. The seaward limit is approximately at the river mouth bar; the upper limit is probably the limit of the salt-water wedge.

INTERDISTRIBUTARY BAY FAUNA

This fauna occurs between the natural levees of river distributaries and is characterized by abundant *Elphidium delicatulum* Bermúdez. The following also are important: *Ammotium salsum* (Cushman and Brönnimann) and vars., *Elphidium gunteri* Cole, *Miliammina fusca* (Brady), and *Streblus beccarii* (Linné) vars. (Lankford, 1959).

DELTAIC MARINE FAUNA

In the Mississippi Delta this fauna occurs seaward from the active passes where there is rapid deposition of sediment and where the environment is essentially open ocean (Lankford, 1959). The assemblage is characterized by *Bolivina lowmani* Phleger and Parker, *Buliminella* cf. *B. bassendorfensis* Cushman and Parker, *Epistominella vitrea* Parker, and *Nonionella opima* Cushman. These four species are often 90% of the population. Living populations are the highest yet found, but below the surface sediment the abundance of dead forms is diluted by rapid deposition of river-borne detritus. Specimens are small in size.

A fauna at the mouth of the Guadalupe River has general features similar to Lankford's deltaic marine fauna of the Mississippi Delta. The Guadalupe River assemblage has very large living populations of two or three species in an area of relatively rapid deposition. The species composition is different, however; the fauna is dominated by *Ammotium salsum*

(Cushman and Brönnimann) vars., and also contains abundant *Palmerinella gardenislandensis* (Akers), *Streblus beccarii* (Linné) vars., and *Elphidium* spp. Specimens are smaller than the average size of the same species found in other environments. Here the river water enters a lagoon instead of the open ocean.

It is believed that the deltaic marine fauna is an important one which can be recognized by 1) large living populations of very few species, and 2) small size of specimens. The species composition varies with the faunal composition of the area into which the river empties. Beneath the surface of the sediment the populations may be low due to rapid deposition.

Other Studies on Nearshore Distributions

EUROPEAN AREAS

Kruit (1955), in his analysis of the microfaunas of the Rhone Delta, has distinguished the following environments with their characteristic Foraminifera:

A. Terrestrial environments
   1. Freshwater–oligohaline basin (permanently swampy)
      Haplophragmoides canariensis d'Orbigny var. *provencensis* Kruit
   2. Oligohaline-mesohaline basin (occasionally dry)
      Haplophragmoides canariensis d'Orbigny var. *provencensis* Kruit
      Trochammina inflata (Montagu)
      Rotalia beccarii (Linné) [=*Streblus*]
   3. Oligohaline-mesohaline basin (permanently swampy)
      Haplophragmoides canariensis d'Orbigny var. *provencensis* Kruit
      Trochammina inflata (Montagu)
      Rotalia beccarii (Linné) [=*Streblus*]
      Nonion cf. *depressulum* (Walker and Jacob)

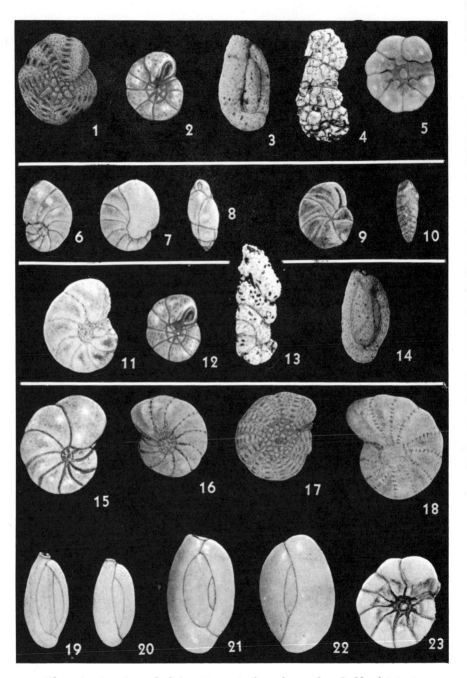

Plate 9. Beach and delta Foraminifera from the Gulf of Mexico.

EXPLANATION OF PLATE 9

Fluvial Marine
1. *Elphidium gunteri* Cole. × 58.
2. *Palmerinella gardenislandensis* (Akers). × 85.
3. *Miliammina fusca* (Brady). × 45.
4. *Ammotium salsum* (Cushman and Brönnimann). × 50.
5. *Streblus beccarii* (Linné). × 62.

Deltaic Marine
6, 7. *Nonionella opima* Cushman. × 70.
8. *Buliminella* cf. *B. bassendorfensis* Cushman and Parker. × 70.
9. *Epistominella vitrea* Parker. × 70.
10. *Bolivina lowmani* Phleger and Parker. × 62.

Interdistributary Bay
11. *Elphidium delicatulum* Bermudez. × 70.
12. *Palmerinella gardenislandensis* (Akers). × 85.
13. *Ammotium salsum* (Cushman and Brönnimann). × 50.
14. *Miliammina fusca* (Brady). × 45.

Beach
15. *Hanzawaia strattoni* (Applin). × 50.
16. *Elphidium incertum mexicanum* Kornfeld. × 70.
17. *Elphidium gunteri* Cole. × 38.
18. *Elphidium discoidale* (d'Orbigny). × 50.
19, 20. *Quinqueloculina compta* Cushman. × 50.
21, 22. *Quinqueloculina seminulum* (Linné). × 50.
23. *Streblus beccarii* (Linné) variant A. × 38.

4. Oligohaline-mesohaline basin (open lake)
   Rotalia beccarii (Linné) [=*Streblus*]
   Nonion cf. *depressulum* (Walker and Jacob)
   Elphidium lidoense Cushman var. *camarguensis* Kruit
   E. crispum (Linné)
5. Mesohaline–strongly saline basin
   Rotalia beccarii (Linné) [=*Streblus*]
   Nonion cf. *depressulum* (Walker and Jacob)
   Elphidium lidoense Cushman var. *camarguensis* Kruit
6. Polyhaline–strongly saline basin. Same fauna as no. 5 and also
   Elphidium crispum (Linné) var.
   Quinqueloculina cf. *elegans* d'Orbigny var.

B. Nearshore marine environments
   1. Restricted to active transport zone to approximately 20 m.:
      Discorbis globularis (d'Orbigny) var. *bradyi* Cushman
      Nonion asterizans (Fichtel and Moll)
      Quinqueloculina rugosa d'Orbigny
      Rotalia faramanensis Kruit
      Triloculina cf. *bermudezi* Acosta
      T. inflata d'Orbigny
   2. Restricted to quiet zones of finer sediments:
      Elphidium lidoense Cushman
      Haplophragmoides glomeratum (Brady) [=Adercotryma]
      Planorbulina mediterranensis d'Orbigny
      Proteonina difflugiformis (Brady)
      Pyrgo inornata (d'Orbigny)
      Quinqueloculina cf. *Q. suborbicularis* d'Orbigny
      Reophax cylindrica Brady
      Textularia agglutinans d'Orbigny
      Triloculina longirostra (d'Orbigny) var. *soldanii* (d'Orbigny)
      T. longirostra (d'Orbigny) var. *italica* (Terquem)
      Verneuilina scabra (Williamson)

Bartenstein and Brand (1938) in their study of benthonic Foraminifera in Jade Bay, in the North Sea, recognized a marine assemblage and a brackish assemblage. The marine assemblage contained:

>*Elphidium excavatum* (Terquem)
>*Globigerina bulloides* d'Orbigny
>*Nonion depressulum* (Walker and Jacob)
>*Quinqueloculina seminulum* (Linné)
>*Reophax nodulosus* Brady
>*Rotalia beccarii* (Linné) [=*Streblus*]
>*Triloculina oblonga* (Montagu)
>*Verneuilina scabra* (Williamson)

The brackish-water fauna contained:

>*Ammobaculites agglutinans* (d'Orbigny)
>*Bigenerina nodosaria* d'Orbigny
>*Haplophragmoides canariensis* (d'Orbigny)
>*Jadammina polystoma* Bartenstein and Brand
>*Proteonina difflugiformis* (Brady)
>*P. fusiformis* Williamson
>*Quinqueloculina arenacea* (Rhumbler)
>*Q. fusca* Brady
>*Trochammina inflata* (Montagu)
>*T. nitida* Brady

Rottgardt (1952) recognized the following environments and their distinctive Foraminifera in the Kiel Bay area on the coast of Germany:

Marine (20-30 o/oo salinity)

>*Eggerella scabra* (Williamson)

Brackish-marine (9-20 o/oo salinity)

>*Reophax subfusiformis* Earland
>*R. dentaliniformis* Brady
>*Elphidium* sp.

Brackish (6-9 o/oo salinity)

> *Miliammina fusca* (Brady)
> *Elphidium granulosum* Galloway and Wissler
> *E. asklundi* Brotzen
> *Nonion depressulum* (Walker and Jacob)
> *Trochammina nitida* Brady

Brackish lake (1.8-6 o/oo salinity)

> *Miliammina fusca* (Brady)
> *Proteonina difflugiformis* (Brady)

LeCalvez and LeCalvez (1951) studied several samples from two bays along the coast of France. The Foraminifera which they reported are typical bay and/or nearshore assemblages of calcareous types. Blanc-Vernet (1958) in western Provence distinguished shell beach, lagoon, vegetated nearshore areas, and sandy and muddy bottom beyond.

Van Voorthuysen (1951) has reported on the Recent Foraminifera in forty-six bottom samples from the Netherlands Wadden Sea, a tidal flat area with numerous tidal channels inland from the Frisian Islands. He records fifty-eight species, of which the following comprise more than 95% of the population:

> *Rotalia beccarii* (Linné) [=*Streblus*]–52.2%
> *Elphidium excavatum* (Terquem)
> *E. incertum* (Williamson) var. *clavatum* Cushman
> *Nonion depressulum* (Walker and Jacob)

His fauna seems to be a typical nearshore, open-ocean assemblage. Two marsh species appear to "contaminate" the fauna: *Trochammina inflata* (Montagu) and *T. macrescens* H. B. Brady.

Pratje (1931) found the following forms dominant in Helgoland Bay, north of the Wadden Sea:

> *Rotalia beccarii* (Linné) [=*Streblus*]—52.5%
> *Nonion depressulum* (Walker and Jacob)—15.5%
> *Polystomella striatopunctata* (Fichtel and Moll)—9.3%
> *Globigerina bulloides* d'Orbigny—2.9%

This also is a typical open-ocean, nearshore assemblage.

SOUTH AMERICA

Hedberg (1934) recognized in the Maracaibo Lake region of Venezuela and elsewhere the basic distributions in some brackish and nearshore marine species, as follows (*op. cit.*, p. 475): "*Quinqueloculina fusca* Brady [= *Miliammina*] is commonly a brackish to fresh water species. . . . *Rotalia beccarii* (Linné) [= *Streblus*] is a brackish to marine species. . . . In shallow (marine) water it is commonly associated with species of *Elphidium, Nonion, Textularia, Cibicides* and various miliolid genera. When occurring alone or with *Quinqueloculina fusca*, brackish water is indicated. . . . *Haplophragmoides, Ammobaculites, Trochammina* and perhaps other arenaceous genera have had representatives in brackish water as well as in truly marine environments."

Van Andel *et al.* (1954) have made a comprehensive study of the Gulf of Paria, between the Orinoco delta coast of Venezuela and the island of Trinidad, B.W.I., and Kruit (*op. cit.*) has given the distribution of Foraminifera in this area. The fauna in the central part of the Gulf of Paria is dominated by *Nonionella atlantica* Cushman and "*Rotalia*" *rolshauseni* Cushman and Bermúdez. The following also occur in significant frequencies:

> *Virgulina pontoni* Cushman
> *Elphidium poeyanum* (d'Orbigny)
> *E. discoidale* (d'Orbigny)

There is a zone near the deltaic shore of the mainland where no microfauna is reported. Between this sterile zone and the *Nonionella atlantica*-"*Rotalia*" *rolshauseni* zone is an area where specimens are unusually small and occur in very small quantities. The following are present:

> *Bolivina* cf. *B. barbata* Phleger and Parker
> *B. striatula* Cushman var. *spinata* Cushman
> *Elphidium poeyanum* (d'Orbigny)
> *Quinqueloculina lamarckiana* d'Orbigny
> "*Rotalia*" *rolshauseni* Cushman and Bermúdez
> *R. beccarii* (Linné) [=*Streblus*]
> *Virgulina pontoni* Cushman

The Bocas del Dragon, a deep-water area between Peninsula de Paria and Trinidad, contains *Bulimina marginata, B. pupoides, Cibicides* cf. *C. concentricus,* and *Textularia pseudogramen,* in addition to *Nonionella atlantica* and "*Rotalia*" *rolshauseni* from the bay.

Numerous other faunal details given by Kruit may be obtained from the original paper. The general pattern is somewhat similar to nearshore distributions obtained from the northern Gulf of Mexico and elsewhere.

Todd and Brönnimann (1957) have studied the nearshore foraminiferal assemblages from the eastern Gulf of Paria on the west coast of Trinidad, B.W.I. Some general features of the different faunas are shown in Figure 57a-d, and the comments of the authors (*op. cit.*, pp. 6-7) are as follows:

"The arenaceous portion of the tidal assemblages (Mangrove I, II) is much larger than that of the nearshore and offshore zones. In Mangrove II more than 90 percent of the specimens fall in arenaceous species, whereas Mangrove I, which is in part from stations transitional to the nearshore zone, has only about 55 percent of arenaceous specimens. The arenaceous specimens of the nearshore zone [0-2 fm.] constitute about 35

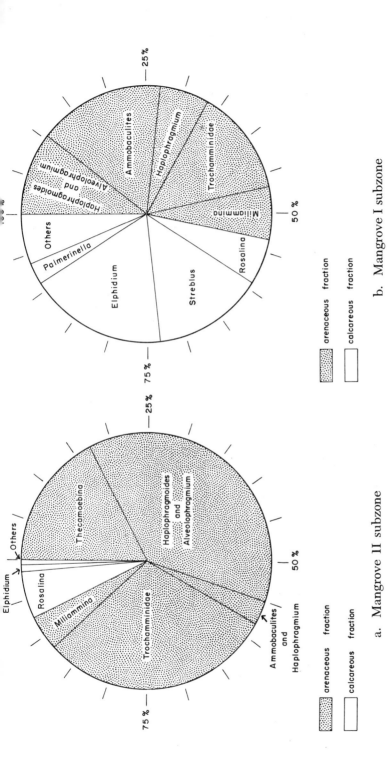

a. Mangrove II subzone

b. Mangrove I subzone

Figure 57. Composition of composite microfaunal assemblage in the eastern Gulf of Paria. After Todd and Brönnimann (1957).

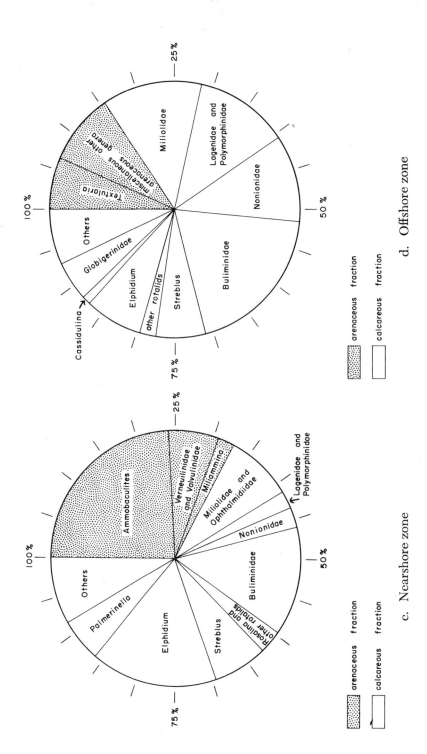

Fig. 57. Continued

c. Nearshore zone
d. Offshore zone

per cent of the microfauna; that of the offshore zone about 20 per cent. From the tidal areas to the deeper waters of the Gulf of Paria, the number of arenaceous specimens decreases, whereas that of the calcareous ones increases. . . . The composition of the arenaceous groups changes from a *Trochammina—Ammobaculites—Haplophragmium—Haplophragmoides —Alveolophragmium—Miliammina* association in Mangrove I, to an *Ammobaculites—Eggerella—*Verneuilinidae*—Miliammina* association in the nearshore zone, to a predominantly *Textularia* association in the offshore zone [2-18 fm.]. The calcareous group passes from an *Elphidium—Streblus—Rosalina—Palmerinella* association in Mangrove I to a Buliminidae—Miliolidae—Ophthalmidiidae—*Elphidium—Streblus—Palmerinella* association in the nearshore zone to a Buliminidae—Nonionidae—Lagenidae—Polymorphinidae—Miliolidae—*Elphidium—Streblus* —Globigerinidae association in the offshore zone [2-18 fm.]. The main faunal break in the arenaceous group occurs between the nearshore and the deeper water zones. The calcareous species show distinct faunal breaks between each of the three zones. . . ."

Drooger and Kaasschieter (1958) have suggested the following foraminiferal zones on the Orinoco-Trinidad-Paria shelf based on the dead populations:
1. The coastal zone of the Orinoco shelf contains *Streblus beccarii* (Linné) and *Streblus sarmientoi* (Redmond) in silty and sandy pelite.
2. In the inner pelitic area of the Orinoco shelf there is a *Nonionella-Uvigerina* zone containing *Nonionella atlantica* Cushman, *Cancris sagra* (d'Orbigny), *Uvigerina peregrina* Cushman, and *Virgulina pontoni* Cushman in some abundance.
3. A *Uvigerina* zone occurs in a pelite area north of Trinidad-Paria containing a high percent of *Uvigerina peregrina* Cushman and also having numerous *Höglundina elegans* (d'Orbigny), *Eponides regularis* Phleger and Parker and

Uvigerina proboscidea Schwager.

4. A shallow reef area east of Trinidad contains abundant *Amphistegina lessonii* d'Orbigny, *Cibicides pseudoungerianus* (Cushman), Miliolids, and arenaceous species.

Boltovskoy (1954c) studied the Foraminifera in nine cores from Golfo San Jorge in southern Argentina. He reports the fauna as being "monotonous and poor," with only a few abundant species and most forms very rare. Many of the species are very small in size, in many a partial or complete loss of ornamentation occurs, and there are other irregularities in form. Poor food supply is suggested as a possible reason for the depauperate nature of the faunas. This interpretation is in agreement with observations of Bradshaw (personal communication) who has seen similar effects in experimental cultures of Foraminifera. Other experimentally observed causes of test deformity or small size are extreme temperatures and salinities. Boltovskoy (1957) also has described the faunas of the Rio de la Plata in Argentina, and Recent Foraminifera from nearshore areas in southern Brazil (1959b).

ASIA

Matsukawa-Ura Bay in Japan is a lagoon connected with the open ocean by a narrow inlet and characterized by lower-than-oceanic salinities which are variable (Takayanagi, 1955). The lagoon fauna is composed of more than 80% arenaceous types, a low population, and relatively few species. The principal forms in this assemblage are as follows:

>   *Haplophragmoides canariensis* (d'Orbigny)
>   *Ammobaculites agglutinans* (d'Orbigny)
>   *Goësella iizkae* Takayanagi
>   *Miliammina fusca* (Brady)
>   *Trochammina globigeriniformis* (Parker and Jones)
>   *T. inflata* (Montagu)

T. cf. T. *nanus* (Brady)
Elphidium matzukawauraense Takayanagi
Rotalia beccarii (Linné) [=*Streblus*]

The open-ocean fauna is composed of almost entirely calcareous types, contains many more species and has a larger population. In and near the inlet there is mixing between the open-ocean and lagoon faunas with the assemblage essentially like that of the open ocean.

Two assemblages have been described from Tokyo Bay by Morishima (1955). Characteristic species of the lower bay are:

Miliolidae
Cassidulina
Cibicides
Eponides
Textularia conica d'Orbigny
T. abbreviata d'Orbigny
Amphistegina radiata Terquem

The inner bay is dominated by *Trochammina globigeriniformis* (Parker and Jones), which here comprises more than 80% of the population, and also contains:

Rotalia beccarii (Linné) [=*Streblus*]
Textularia parvula Cushman
Proteonina difflugiformis (Brady)
Elphidium fabum (Fichtel and Moll)
Nonionella miocenica stella Cushman and Moyer
Nonion manpukujiensis Otuka

In the inner bay these forms are small and thin-walled.

Morishima (1948) has recorded the following asemblages from Ago Bay, Japan:
1. *Globigerina* assemblage at the entrance to the bay.
2. *Amphistegina–Elphidium crispum* assemblage is near the

mouth of the bay. *Rotalia rosea* and *Cibicides refulgens* also occur.

3. *Rotalia papillosa—Quinqueloculina lamarckiana* assemblage is in the center portion of the main bay.

4. *Elphidium craticulatum—Textularia hauerii* assemblage occurs in certain sections of the inlets.

5. *Trochammina* assemblage is characteristic of the innermost bay and also contains *Haplophragmoides* and a small percentage of Nonionidae.

Morishima and Chiji (1952) have studied the distribution of Foraminifera in Akkeshi Bay, Japan, and have delineated five assemblages in different parts of the bay, characterized as follows:

1. The *Trochammina globigeriniformis* (Parker and Jones) —"*Rotalia*" *beccarii* (Linné) assemblage occurs in Akkeshi Lake which branches off from the main bay and is separated from it by a narrow inlet. Other characteristic species are:

> *Eponides frigidus* Cushman [=*Buccella*]
> *Elphidium incertum* (Williamson)
> *Buliminella elegantissima tenuis* Cushman and McCulloch
> *Elphidium* cf. *E. etigoense* Husezima and Maruhasi

2. Near the inlet connecting the two bays the fauna is almost entirely *Trochammina globigeriniformis* (Parker and Jones).

3. The inner part of the main bay has the *Bolivina decussata* (Brady) assemblage which also contains common *Elphidium advena* (Cushman) and *Pseudononion japonicum* Asano.

4. The fauna in the major part of the center of the bay and toward the mouth is dominated by *Rotalia japonica* Hada, *Pseudononion japonicum* Asano and *Nonionella pulchella* Hada. Common additional species are *Eponides schreibersii* (Reuss), *Nonion nicobarense* Cushman, *Cibicides lobatulus* (Walker and Jacob) and *Elphidiella* cf. *E. sibirica* (Goës).

LeRoy (1938, p. 130) in his study of samples from a traverse across Peper Bay, on the west coast of Java, noted: "Three rather well-defined communities have been observed . . . : (1) *Haplophragmoides-Haplophragmium* (semi-brackish facies); (2) *Operculina-Ozawaia* (open-sea facies); (3) *Dendritina-Alveolinella* (protected shoal facies). These three faunal units are quite distinctive and each is characterized by several species which appear to be confined to their respective habitats. Salinity and relationship of bay areas to open sea conditions appear to be the major controlling factors in determining the distribution of the faunas."

HYPERSALINE COASTAL LAGOONS

Foraminifera assemblages are described from two markedly hypersaline lagoons in the Pacific, Shark Bay, Western Australia (Logan, 1959), and Laguna Ojo de Liebre, Baja California, Mexico (Phleger and Ewing, in press). Both these lagoons have inner lagoon and lower lagoon faunas. Faunal composition differs and they both differ from the Gulf Coast lagoons, but the faunal pattern is similar. The development of an inner lagoon fauna may be dependent in part on the size of a lagoon, with an ecologically distinctive body of water developing at some distance from any outlet to the open ocean.

Discussion of Marsh Faunas

One of the surprising features of the marsh fauna is that many of the species seem to have a very wide geographic distribution. *Ammoastuta inepta* (Cushman and McCulloch) may be used as an example of a form which may be worldwide in distribution; it is reported from Trinidad, California, northern Gulf of Mexico, northeastern United States, Panama, and Ecuador, and probably is present in marshes elsewhere. It is difficult to imagine the mechanism by which this form has been distributed so widely since it appears to be restricted to a marine marsh environment. Marshes do not continuously fringe coasts, nor are the coasts themselves continuous, as would seem to be required for the easy dispersal of such a form. Such a distribution may have become established over a long period of time when coastal marine marshes were much more extensive than at present. It is possible, for example, that coastal marshes were more widespread at some time during mid-Tertiary than they are at present.

Although the faunas of marine marshes are generally similar in those localities where they have been studied, there are somewhat different marsh assemblages. This is illustrated by the marsh faunas in the San Antonio Bay area on the coast of Texas and summarized by F. L. Parker *et al.* (1953, p. 16) as follows:

"Two small areas of marsh have been studied from this area, one on the bay side of Matagorda Island and the other at Grassy Point on the delta of the Guadalupe River. Each of these marshes has a somewhat different ecology and there are considerable faunal differences between some stations in the same marsh. A few species are essentially restricted to these marsh environments, with some mixing in adjacent areas. Such species found in both marshes are:

> *Ammoastuta inepta*
> *Arenoparrella mexicana*
> *Discorinopsis aguayoi*
> *Miliammina fusca*
> *Trochammina comprimata* [=*Tiphotrocha*]
> *T. inflata*
> *T. macrescens*

"The following species are confined to the Matagorda Island marsh:

Jadammina polystoma
Pseudoclavulina gracilis
Triloculina fiterrei meningoi

... *Trochammina inflata* is much more abundant in this marsh than at Grassy Point. The marsh is characterized by relatively high populations of *Palmerinella palmerae* and also contains several species which are characteristic of both the bay facies and the open-gulf facies. *Massilina protea* is more abundant here than elsewhere.

"*Miliammina* sp. is restricted to the Grassy Point marsh. *Discorinopsis aguayoi* occurs here at very high frequencies and *Trochammina* [= *Tiphotrocha*] *comprimata* is more abundant than on Matagorda Island."

Marsh species which occur in estuaries generally are present in much lower frequencies than in the marsh. Very narrow estuaries probably cannot be distinguished faunally from the marsh. In the Barnstable area (Phleger and Walton, 1950, p. 283) the following zones contain distinctive Foraminifera: *Spartina patens* zone, *S. glabra* zone, intertidal flats, and the *Zostera* zone.

One of the characteristics of marsh foraminiferal assemblages is that many of them are largely or entirely composed of species having arenaceous tests. This may be related to a low pH in the marsh environment. Under these conditions it seems probable that calcareous tests cannot survive regardless of other environmental factors.

Some data are available on the hydrogen-ion concentration in the sediment of marshes. In Newport Bay, California, Emery and Stevenson (1957) report that 60% of the marsh sediment had a pH of more than 7.0. Average values for pH given by them for the floral zones found in the Newport Bay marsh are: *Zostera* zone 7.5, *Spartina* zone 7.2, *Salicornia* zone 6.8, and *Distichlis* zone 5.9.

A small area of marsh has been studied in the Baptiste Collette Subdelta of the Mississippi Delta. Values of pH in the sediment obtained here range from 6.79 to 7.51 and comparable values were obtained in the rivers and nearshore interdistributary bays. In upper San Antonio Bay, Texas, values of pH decrease from 7.37 to 6.61 toward the delta of the Guadalupe River with its marshes.

These meagre data suggest that some marine or brackish marsh sediments have a pH of less than 7.0. Most of the measurements probably were taken during the daylight hours when photosynthesis of the dense plant population was using carbon dioxide from the water and tending to raise the pH. At night the reverse occurs, with carbon dioxide being added to the water by plant respiration and a resulting lowering of pH. Stevenson (1954) found the highest pH in Newport Bay, California, in the afternoon and the lowest before sunrise. Moberg (1927) found a diurnal range of 0.25 in pH units in the water of Mission Bay, California, with the lowest just before sunrise. These observations indicate that marsh sediments may have a considerably lower pH value at night than during the day and these low night values may be a critical factor in some foraminiferal distributions where calcareous tests are not found.

Two recent marsh studies are of interest. Behm and Grekulinski (1958) have examined marsh and small estuaries from Staten Island. They conclude that the Foraminifera are commensal with the marsh floras. Parker and Athearn (1959) have examined faunas from Poponessett Bay, Massachusetts, where the marsh varies from weakly brackish to essentially pure marine. They find that the following decrease in abundance with increasingly marine conditions: *Arenoparrella mexicana* (Kornfeld), *Haplophragmoides hancocki* Cushman and McCulloch, *Tiphotrocha comprimata* (Cushman and Brönnimann) and *Trochammina macrescens* Brady. *Jadammina polystoma* Bartenstein and Brand and *Trocham-*

mina inflata (Montagu), on the other hand, increase in abundance with increasing salinity. Several other species seem to show no correlation with salinity.

THECAMOEBINA

Bolli and Saunders (1954) have indicated that the following forms should not be classed with the Foraminifera, but should be listed with the Thecamoebina, largely under the generic names *Centropyxis* and *Difflugia: Leptodermella* (part), *Urnulina* (part), *Lagunculina* (part), *Proteonina lagenaria, P. hancocki,* and others. Such forms appear to be largely, or perhaps exclusively, fresh-water in habitat. These authors have studied the distribution of Thecamoebina in the rivers of Trinidad, B.W.I., and have established the distribution of four "zones," as follows (*op. cit.,* pp. 45-46):

"Zone I. Swiftly flowing rivers with pools and small waterfalls in mountainous rain forest. Water clear and well aerated. Characterized by the abundance of the genus *Centropyxis,* especially *Centropyxis (Cyclopyxis) stellata* Wailes, *Centropyxis (Centropyxis) ecornis* Leidy and *Centropyxis (Centropyxis) aculeata* Stein and certain species of the genus *Difflugia,* e.g., *D. capreolata* Penard and *D. caronia* Wallich.

"Zone II. Rivers flowing in the upper part of the zone in the forest and lower down in cultivated land. Water clear to slightly turbid, and though flowing appreciably not as fast as in Zone I. Characterized by the abundance of the genus *Difflugia,* especially common being *D. urceolata* Carter. The genus *Centropyxis* is reduced in numbers in this zone, those remaining belonging mostly to the subgenus *Centropyxis.*

"Zone III. Rivers flowing through cultivated land. Water turbid, slow flowing or semi-stagnant. Thecamoebian fauna rather impoverished. Best represented is the genus *Difflugia,* especially a form near to *D.* cf. *acuminata* Ehrenberg.

"Zone IV. Rivers bordered by swamp vegetation (mangroves, etc.). Water brackish and turbid, slow flowing or semi-stagnant. Exceedingly impoverished fauna. The specimens collected were not studied for protoplasm but it is the authors' opinion that they are more likely to have been carried down than be living there."

Bolli and Saunders believe that these forms probably are not preserved as fossils, and that reports of fossil occurrences are due to contamination at collecting locations. It is the experience of the writer that these forms often are abundant in Recent (modern) samples which have never been dried out, and that many or most specimens disintegrate as a result of the drying process. It is possible that such forms do survive as fossils in certain silts or clays and might be recovered if proper care were exercised in preparation of the samples for study.

"*Leptodermella variabilis*" [*Centropyxis*] and "*Urnulina*" [*Difflugia*] are characteristic of the innermost coastal marsh in the Mississippi Sound area, in the northern Gulf of Mexico (Phleger, 1954a). This inner marsh contains very weakly brackish or fresh water, and the occurrences of these species probably are explained on the basis of the distributions analyzed by Bolli and Saunders (1954).

In the San Antonio Bay area in the northern Gulf of Mexico, F. L. Parker *et al.* (1953) have recorded "*Leptodermella variabilis*" [*Centropyxis*], "*Proteonina lagenaria*" [*Difflugia*] and "*Urnulina compressa*" [*Centropyxis*] in the Guadalupe River and in the upper bay sediments near the mouth of the river. F. L. Parker (1952) records several of these forms in the lower Housatonic and Connecticut rivers and near the mouths of these rivers in Long Island Sound. It is apparent in such occurrences that the Thecamoebina have been carried from the rivers and their presence in the bay or lagoon sediments reflects the presence, therefore, of such rivers. Lowman (1949, p. 1954) recorded *Difflugia* and *Centropyxis* in the Mississippi Delta and correctly interpreted them as representing fresh and weakly brackish water environments.

Coral Reef Faunas

Cushman *et al.* (1954) published an excellent paper on the distribution of Foraminifera in Rongerik, Rongelap, Bikini, and Eniwetok atolls of the Marshall Islands in the central Pacific. This is one of several papers describing and interpreting the oceanography and geology of these islands. The authors have listed faunas typical for reef flat, beach, lagoon, outer slope, and deep water.

The reef flat is dominated by *Calcarina spengleri* (Gmelin), *Marginopora vertebralis* Blainville, *Homotrema rubrum* (Lamarck), *Miniacina miniacea* (Pallas), and *Carpenteria proteiformis* Goës. Small percentages of *Amphistegina madagascarensis* d'Orbigny occur, and approximately thirty other species are present in low frequencies mainly in tide pools. Occasional planktonic Foraminifera are present.

The beach fauna consists mainly of worn specimens. The composition of a typical beach sample is given as (per cents approximate):

> *Amphistegina madagascarensis* d'Orbigny—55%
> *Marginopora vertebralis* Blainville—25%
> *Calcarina spengleri* (Gmelin)—15%
> Other Foraminifera 5%

The lagoon fauna is dominated by *Amphistegina madagascarensis* d'Orbigny and occasionally by *Heterostegina suborbicularis* d'Orbigny. These and the following comprise 50-75% of the lagoon assemblage:

> *Calcarina hispida* Brady
> *C. spengleri* (Gmelin)

Homotrema rubrum (Lamarck)
Marginopora vertebralis Blainville
Miniacina miniacea (Pallas)

Approximately 185 species comprise the remaining 25-50% of the lagoon fauna.

The composition of a typical lagoon sample is given as (per cents approximate):

Amphistegina madagascarensis d'Orbigny—5%
Heterostegina suborbicularis d'Orbigny—10%
Marginopora vertebralis Blainville—5%
Calcarina hispida Brady—5%
Homotrema and *Miniacina* 5%
Buliminidae 5%
Miliolidae 6%
Nonionidae 2%
Textulariidae 6%
Other Foraminifera 6%

The following species are essentially restricted to the lagoons:

Bulimina fijiensis Cushman
Buliminella milletti Cushman
Discorbis subbertheloti Cushman
Epistomaroides polystomelloides (Parker and Jones)
Gypsina plana (Carter)
Hauerina serrata Cushman
Monalysidium politum Cushman
Nubecularia lacunensis Chapman
Patellinella inconspicua (Brady)
Quinqueloculina anginna var. *arenata* Said
Reussella sp. A
Rotalia cf. *R. beccarii* var. *tepida* Cushman [=*Streblus*]
Rugidia spinosa Cushman
Siphonina tubulosa Cushman
Spirillina decorata Brady
S. vivipara var. *revertens* Rhumbler

Spiroloculina clara var. *lirata* Cushman
S. foveolata Egger
Triloculina sp. A

Globigerinoides ruber (d'Orbigny) and *G. sacculifer* (Brady) are found throughout the lagoons. These are the two most abundant planktonic species in the region.

According to the authors (*op. cit.*, p. 321), "The samples from the outer slopes of the atolls do not differ greatly in precentage composition from those inside the lagoons. They are, however, characterized by the appearance, as rare specimens, of various deep-water species that occur in greater abundance at greater depths." The following species occur only on the outer slopes:

Siphotextularia crispata (Brady)
Gaudryina (Siphogaudryina) siphonifera (Brady)
Cycloclypeus carpenteri Brady
Discorbis tuberocapitata (Chapman)
Cassidulina pacifica Cushman

The "deep-water" fauna (deeper than 55 fm. [101 m.]) is dominated by planktonic forms (50-98%) and about 215 benthonic forms appear. At the shoaler locations species of *Amphistegina* may be abundant.

Foraminifera are an important constituent of the sandy sediment in coral reefs. This is well-illustrated by the description of the sedimentary facies at Kapingamarangi Atoll (McKee *et al.*, 1959) which extend from the reef to the center of the lagoon at 240 ft. In shallow water (less than 6 ft.) *Amphistegina madagascarensis* d'Orbigny makes up one-half of the volume of the sandy sediment. At depths of 156-210 ft. *Amphistegina lessonii* d'Orbigny is 85% of the sediment.

W. E. Moore (1957) has studied foraminiferal distributions in 16 samples from the northern Florida Keys. He has separated the area into four environments which he recognizes mainly on the occurrences of families of Foraminifera, as follows (*op. cit.*, p. 732):

"1. Florida Bay environment west of the Florida Keys, water 0-10 ft. [0-30 m.] deep. The foraminiferal families Miliolidae, Peneroplidae, and Nonionidae are dominant. *Rotalia beccarii* and *Cornuspiramia antillarum* are restricted to this environment.

"2. The back-reef environment lies between the Florida Keys and the reef at the east, with water mostly 20-30 ft. [6-9 m.] deep. The foraminiferal families Miliolidae, Peneroplidae, Nonionidae and Rotaliidae dominate the back-reef fauna. The fauna is composed of more families, genera, and species than that of Florida Bay. The families Camerinidae and Alveolinellidae, although very rare, appear to be characteristic of this environment. . . .

"3. The reef environment lies east of the Florida Keys and is locally awash at low tide. . . . The Peneroplidae are the dominant Foraminifera . . . and the Amphisteginidae, *Asterigerina* especially, are characteristic. The Nonionidae are at a minimum near the outer reef patches.

"4. The fore-reef environment extends seaward into the Atlantic from the reef. A number of gradational environments are present there and they appear to be generally comparable with the deeper-water, open-ocean environments proposed by other workers."

Illing (1952) reports distributions of Foraminifera in 19 samples from the Bahama Banks. She records the following species as restricted to or more common in exposed localities at the open-ocean edge of the banks:

Amphistegina lessonii d'Orbigny
Bigenerina nodosaria d'Orbigny var. *textularioidea* (Goës)
Dentostomina spp.

Eponides antillarum (d'Orbigny)
E. repandus (Fichtel and Moll)
Heterostegina antillarum d'Orbigny
Homotrema rubrum (Lamarck)
Quinqueloculina bradyana d'Orbigny
Q. lamarckiana d'Orbigny
Q. tricarinata d'Orbigny
Rotalia rosea d'Orbigny
Textularia agglutinans d'Orbigny
Triloculina trigonula (Lamarck)

Articulina mexicana Cushman is restricted to sheltered waters, and other forms which are common in the shallow water of the banks are various species of Peneroplidae, Miliolidae, and Rotaliidae. *Elphidium morenoi* Bermúdez and *Streblus beccarii* (Linné) are restricted to brackish waters. Zones of mixed faunas occur in and around the channels where there are strong tidal currents.

CHAPTER IV

Distribution Studies of Living Benthonic Foraminifera

General Discussion

Many of the distributions described in Chapters II and III are based on the total populations of benthonic Foraminifera, and the living population was not considered. In the delineation of foraminiferal distributions it is desirable to know the distribution of the living population, since the dead population alone may not represent the true pattern of the living population. Moreover, studies of living Foraminifera faunas are a possible key to production rates of these forms. They may be used to estimate relative rates of deposition of the sediment in which they occur, and may provide a clue to "residual" faunas and post-mortem transportation of the tests.

Quantitative data on living benthonic Foraminifera have been obtained from several areas. In the northwestern Gulf of Mexico the living populations were recorded from 217 samples from the continental shelf deeper than about 20 m. and the upper part of the continental slope (Phleger, 1951b). Two hundred eleven samples from the southwestern Gulf of

Maine (Phleger, 1952b) were analyzed; these living-population studies were from the mud-facies samples only. Living benthonic Foraminifera have been recorded from numerous shallow-water stations in the southeastern Mississippi Delta area (Phleger, 1955a; Lankford, 1959). Walton (1955) has made a detailed seasonal study of faunas from Todos Santos Bay in Baja California. F. L. Parker (1954) has analyzed numerous samples collected from the northeastern Gulf of Mexico. Living assemblages have been described from closely-spaced samples on the continental shelf off San Antonio Bay (Phleger, 1956), and Laguna Madre (Phleger, 1960b) in Texas, and seasonal studies have been completed on some of these populations. Living Foraminifera are known from Laguna Ojo de Liebre in Baja California, Mexico (Phleger and Ewing, in press), from the San Diego area (Uchio, 1960), in Santa Monica Bay, California (Reiter, 1959), in Santa Cruz Basin, California (Resig, 1958), in Poponesset Bay, Massachusetts (Parker and Athearn, 1959), in the Oslo Fjord (Christiansen, 1958), and in the Rio de la Plata, Argentina (Boltovskoy, 1958). A few useful generalizations have come from these examinations of living populations, but these are tentative and may be modified by additional work.

One of the interesting problems in distribution of benthonic Foraminifera is whether there has been significant post-mortem transportation of the tests and whether this process has affected the apparent faunal associations. In many of the areas studied, especially lagoons, the distributions of the living populations are quite similar to the distributions of empty tests of the same species. This suggests that there has been no transportation of the living specimens. It is the experience of the writer that specimens containing protoplasm are easily put in suspension in counting dishes by slight agitation of the water. Juvenile forms often have very thin tests and thus have a lower effective density than adult, heavy-shelled forms of the same species. It appears probable that they are transported with little difficulty and this may aid in distributing some species. Small, living benthonic forms occasionally are caught in plankton tows in shallow water offshore from San Diego, and a few specimens have been collected in plankton tows from deep water. If young forms are transported into a foreign environment they may not survive to the adult stage, or they may become adult and not be able to reproduce because of adverse environmental conditions.

In traverses of closely-spaced stations off southwest Texas the dead populations of many species extend deeper than the living populations (Figures 18-21). The following example illustrates these differences in range: living *Streblus beccarii* (Linné) is abundant from the shallow ends of the traverses to approximately 40 m. and is rare to approximately 80 m.; dead forms are abundant to approximately 45 m. and are common to rare to the end of the longest traverse at 110 m. It is possible that these faunas have moved downslope because of waves, water currents, or turbidity currents. It seems more likely, however, that these faunas are residual or relict from a previous environment when sea-level was lower and ecological factors were similar to those at present on the inner shelf.

The "standing crop" of benthonic Foraminifera is the living population existing at any time. This is at least a partial measure of the rate of production, although the standing crop of any group of marine organisms will vary from time to time and may not be directly proportional to production. It is of interest to compare the standing crop of Foraminifera in some of the areas studied. The standing crop of benthonic Foraminifera may be converted into living population per square meter of surface area. In a nearshore area of the southern Gulf of Maine a living population varying from 1000/sq. m. to more than 100,000/sq. m., with an average of approximately 30,000/sq. m., was reported. On the outer shelf of the northwest Gulf of Mexico the population averaged only approximately 10,000/sq. m. (Phleger, 1951b). In the Mississippi Delta the populations apparently vary from approximately 1000/sq. m. to more than 1,000,000/sq. m., with an average of approximately 90,000/sq. m. (Phleger, 1955a). Other samples have been found to contain living populations up to 2,500,000/sq. m. of area.

Myers (1935a) studied the production of *Elphidium crispum* (Linné) in the littoral and sublittoral zones at Plymouth, England, over a 12-month period. He came to the conclusion that up to more than 1000 individuals/sq. ft. of area in the environments at this place were produced each year and that essentially all these were contributed to the sediment.

Additional information is necessary to evaluate these results, and seasonal data especially are required. It seems reasonable that production rates of benthonic Foraminifera should have a definite relationship to the total organic production in any area. Relative production rates of these forms, therefore, should be a key to rates of total organic production.

Relative Rates of Deposition Based on Benthonic Foraminiferal Populations

Marine sediment is composed of materials derived from several inorganic and organic sources. If the rate of addition of one component of the sediment can be determined, a key is provided for determining the rate of deposition. One important component of marine sediments is benthonic Foraminifera, and if the rate of production of these forms can be determined they provide a measure of deposition rates. Reliable data for establishing approximate rates of production of these forms will be laborious and time-consuming to accumulate.

An approximation of *relative* rates of deposition may be obtained from live-total ratios of benthonic Foraminifera.

Abundant living specimens relative to the total population (living plus dead specimens) should indicate relatively rapid addition of sediment other than Foraminifera. Likewise, a small living population relative to the total population would indicate a slow rate of supply of sediment. This is based on the assumption that the rate of production is constant over the period of time that the sediment has been accumulating, and such an assumption is only an approximation. Relative rates of deposition determined by such a method apply only to the surface sediment.

This method has been applied in the southeastern Mississippi Delta area where the total amount of deposition during the period 1860 to 1952 has been compiled by Scruton (1953b, fig. 7) from U. S. Coast and Geodetic Survey soundings. In general, the conclusions about relative rates of sedimentation based on the Foraminifera ratios agree with the relative amounts of sedimentation reported by Scruton. This suggests that the method probably is reliable for differentiating in a general way between areas of slow and rapid deposition. A large number of samples with a reasonable, well-planned distributional pattern is desirable; single samples, or only a few samples, may give misleading results, since there may be considerable population differences between individual samples.

A theoretical procedure for determining relative rates of sedimentation may be summarized briefly as follows:

1. The sediment samples should be of equal area and volume.

2. A census is made of both the living and non-living populations.

3. The living population is at least a partial measure of the rate of addition of the tests of these animals to the sediment.

4. The total population, which includes the living population plus the empty tests, represents the accumulation of

tests of Foraminifera in the sediment during an unknown period of time.

5. The ratio of living to total population at any place will represent the rate of accumulation of the sediment. Comparisons of living-total population ratios will indicate relative rates of deposition in different areas; these ratios may be expressed in per cent for ease of comparison. A high numerical per cent suggests relatively fast deposition of sediment; a low numerical per cent suggests relatively slow deposition.

The following are some qualifications to this approach:

1. The rate of production of benthonic Foraminifera is not known, and the standing crop represents the production for an unknown length of time.

2. Production rates probably vary in different areas and at different times in the same area.

3. The mortality rate is unknown and also probably varies in different places and circumstances and at different times.

4. There may be some loss of tests of Foraminifera after deposition due to solution, abrasion, or reworking of the sediment.

RELATIVE RATES OF SEDIMENTATION IN THE NORTHERN GULF OF MEXICO

Traverse off Mississippi Sound. The frequency distribution of living-total population ratios is shown on Figure 58 for a traverse extending seaward from Horn Island, off the coast of Mississippi, and across the continental shelf to approximately 600 ft. (183 m.) water depth. The striking feature about these ratios is that they are unusually low, being 1% or generally much less than 1%, although the actual living populations are average for samples of this size.

These unusually low percentages of living specimens in the total population suggest that most or all of the area along the offshore part of this traverse is essentially one of non-

192 *Ecology and Distribution of Recent Foraminifera*

deposition at the present time. It may be presumed that the sediment being furnished from the mainland either is being trapped in Mississippi Sound or is being deposited elsewhere in the offshore area, or both.

Mississippi Sound and Mobile Bay. Forty-eight samples from Mississippi Sound and 26 samples from Mobile Bay have been studied. These are spaced along a series of 9 traverses which give relatively good coverage for the area. The frequency of the live-total ratios is given on Figure 58. There is a significant difference between the live-total population ratios given for Mississippi Sound and Mobile Bay and those for the traverse off Horn Island. In the traverse seaward from Horn Island 90% of the ratios (per cents) are less than 1 and none is greater than 2. The material from Mississippi Sound and Mobile Bay, on the other hand, has ratios (per cents) which vary from less than 1 to 100; 74% of the ratios

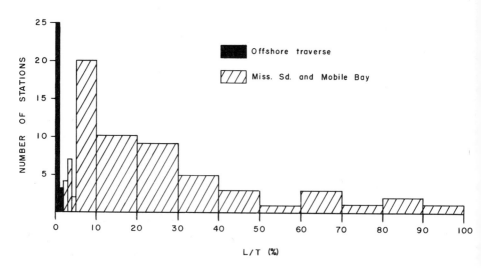

Figure 58. Frequency distribution of live-total ratios of benthonic Foraminifera (in per cent) in Mississippi Sound and Mobile Bay and in offshore traverse from Horn Island.

are 5 or larger, and only 8% are less than 2. Most of the ratios in the sound and bay are between 5 and 39, including 60% of the total, and the mode of the frequency is in the 5-9 group.

These data suggest that the overall rate of deposition in Mississippi Sound and Mobile Bay is significantly greater than in the traverse seaward from Horn Island. The rather wide range in ratio values from place to place may be due to one or all of the following possible causes: (1) local variations in sedimentation rate, (2) variation in quality of the samples as being representative of the area sampled, and (3) local variation in Foraminifera production and/or mortality.

Southeast Mississippi Delta. The live-total population ratios have been calculated for the populations of Foraminifera which have been studied from the southeastern Mississippi Delta area (seen on Figure 59). Distribution of living population in per cent of total population may be divided into four frequency groups which have an areal distribution, indicated on Figure 59 by Roman numerals. Low ratios (I) occur at the westernmost edge of the area sampled in Breton Sound, in samples farthest offshore and near Main Pass. Intermediate areas (II) are in most of Breton Sound and most of the nearshore open gulf with ratios of 10-25%; local high-ratio centers occur in Breton Sound. High ratios (III) occur on the northern side of the inlet between Breton Island and Main Pass and west of Main Pass, and another nearshore area between Main Pass and Pass a Loutre. Areas having considerable range in ratios (IV) are in the marsh of Baptiste Collette Subdelta and the bay between Main Pass and Baptiste Collette Subdelta.

Scruton's (1953b) results indicate rapid deposition near the passes and decrease in deposition offshore, scour at the deepest part of the inlet north of Main Pass, and small and variable depth changes in the inner part of Breton Sound. Relative rates of sedimentation shown by the per cents of living Foraminifera agree with known and inferred sedimen-

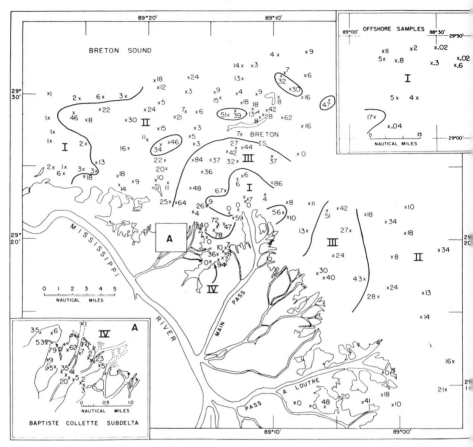

Figure 59. Live-total populations of benthonic Foraminifera in the southeastern Mississippi Delta area, expressed in per cent. After Phleger (1955a).

tation rates in this area. High living frequencies on the northern side of the inlet are exceptions and do not agree with information compiled by Scruton. The cause may be local or short period variation in production or very recent rapid sedimentation. Low sedimentation rate in the offshore samples is striking; this coincides with the results of the Horn Island traverse where low sedimentation is shown in the same area.

San Antonio Bay area. The living-total population ratios in the San Antonio Bay area (Figure 60) fall into two distinct groups. In Aransas Bay, Mesquite Bay, and lower San Antonio Bay the ratios are all 3% or less and mostly 1% or less. The ratios are higher in upper San Antonio Bay, increasing to 11% and 15% as the delta of the Guadalupe River is approached. The total populations are high at most lower bay stations, usually 10,000 or more specimens per 10 ml. of

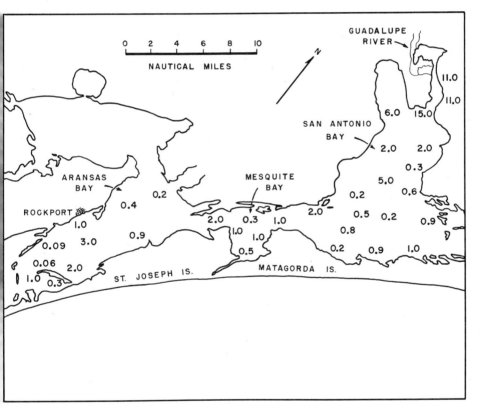

Figure 60. Living-total population ratios of benthonic Foraminifera in San Antonio Bay area, expressed in per cent. After Phleger (1956).

wet sediment; total populations also are high near the delta because of high production of Foraminifera. These ratios suggest that sedimentation rates are relatively high near the delta and low in the lower bays. These conclusions are in essential agreement with those of Shepard (1953), who based sedimentation rates on the amount of shoaling from 1874-1935.

Offshore traverses, northwestern Gulf of Mexico. Ratios between living and total populations have been analyzed from data given in previous publications (Phleger, 1951b, 1960a). These data are from samples along 12 traverses across the continental shelf and upper continental slope in the northwestern Gulf of Mexico between Pt. Isabel and Atchafalaya Bay. Positions of the traverses and summaries of live-total population ratios are on Figure 61.

In most of the traverses there is a marked difference between the living-total population ratios on the inner part of the continental shelf and those at greater depths. The break between these two groups of ratios is distinct, as can be seen by examining the histograms in Figure 61. A relatively sharp break in population ratios occurs in the traverses in this region as follows:

Traverse	Living-total population ratio discontinuity	
	Depth in m.	Approximate distance offshore (nautical miles)
1	40–46	27
2	44–46	32
3	73–79	30
4	48–51	30
5	41–43	28
6	33	27
7	31	28
11	59–66	55
12	64–70	52

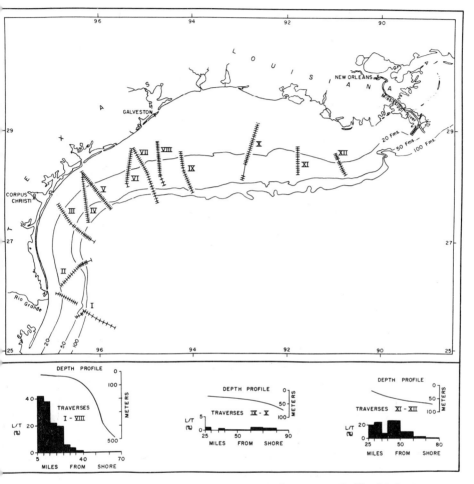

Figure 61. Locations of traverses in northwestern Gulf of Mexico studied for living populations of benthonic Foraminifera. The histograms are live-total population ratios plotted against distance from shore and depth of water in different areas.

These population ratios indicate relatively fast deposition on the inner continental shelf and little or no detrital deposition on the outer shelf. Off Atchafalaya Bay the zone of rapid nearshore deposition extends out approximately 50 mi. to about 65 m. depth. Essentially no sedimentation is indicated on the shelf between Galveston and the Mississippi drainage. On the rest of the Texas shelf there is active sedimentation out for approximately 30 mi. offshore.

The living-total population ratios in several traverses along the shelf near San Antonio Bay (Phleger, 1960a) vary from less than 1% to more than 25%. In general, the ratios are higher nearshore than offshore, but they are quite variable; seaward of approximately 85 m. depth they are significantly lower than nearshore, and there are generally higher ratios shoreward of 30 m. depth. A higher sedimentation rate is indicated in the northwest, probably due to the nearness of the Brazos and Colorado rivers which are sources of supply of sediment for the area.

Northeastern Gulf of Mexico. F. L. Parker (1954) lists the living-total population ratios of benthonic Foraminifera in four traverses from the northeastern Gulf of Mexico. Three of these traverses extend in a southerly direction off the Mississippi Delta and a fourth extends across the continental shelf between Mobile Bay and the southeastern end of the Mississippi Delta. The data along Parker's traverse 4 confirm the data in the Horn Island Traverse, discussed above, in suggesting little active sedimentation on the continental shelf off Mississippi Sound. Relatively high ratios exist off the Mississippi Delta and suggest high relative rates of sedimentation in this area. It is of interest that the high rates are indicated out to a depth of at least 430 m. off Southwest Pass.

Parker (*op. cit.*, p. 475) comments as follows:

"Most of the species present have living representatives to a depth of 200 m.; deeper than this the representation by living forms is spasmodic and they usually form less than 1% of the total population. The various species of *Reophax* are an exception to the general rule. They are usually represented by a relatively large number of living specimens. This is especially true of *R. hispidulus*. The relatively large number of living specimens of these species and the low frequency of dead specimens suggest that these forms are destroyed soon after death. The test of *R. hispidulus* is very fragile, the sand grains which form it being weakly cemented so that once the supporting protoplasm is gone the specimens probably disintegrate rapidly. The same observation was made of such arenaceous species in the shoal samples in the Mississippi Delta region as *Goësella mississippiensis, Nouria polymorphinoides,* and *N.* sp. It is probable that such fragile forms seldom appear in fossil assemblages and are not present in modern dead assemblages in the frequencies warranted by their actual rate of production."

Discussion. The population ratio data from the continental shelf of the northern Gulf of Mexico fit into such a consistent pattern that they suggest a certain reliability. Whether or not the living population represents an average standing crop has to be ascertained by future investigations. It is possible that a very low living population may occasionally exist temporarily and may give an incorrectly low living-total population ratio. The maximum living population in any area may be the most reliable one for these ratios.

It should be pointed out that the methods used in recognizing the living specimens were different in the earlier northwestern Gulf of Mexico samples (Phleger, 1951b) from those used in samples collected later. The Biuret test, used originally, may give lower living population counts than those which actually exist; the rose Bengal test used in other areas is believed to give a fair representation of the actual living populations.

The low population ratios suggest that essentially no deposition of sediment is occurring offshore from Mississippi Sound. If this interpretation is correct, it implies that the sediment being furnished by the Pascagoula River and other streams in the area is being deposited elsewhere. Mississippi Sound and Mobile Bay may act as traps to prevent much or most of the sediment from being deposited seaward from the barrier islands. It is possible also that sediment is being deposited on the seaward side of the barrier islands elsewhere than along the traverses studied.

The surface clastic sediments along the open-ocean part of the Horn Island Traverse may have been deposited during an older cycle of deposition. The modern population of Foraminifera is being supplied to this sediment and is being mixed with faunas which probably were deposited under somewhat different conditions than present ones. These previous environments probably are related to lower sea-levels of late Pleistocene or post-glacial time. The area of apparently low sedimentation rate between Galveston and the Mississippi drainage coincides with a coastline which is free from sandy, barrier islands. This area probably is beyond the sedimentary influences of the Mississippi and Atchafalaya drainage. It is possible that the rivers draining into the area do not carry sufficient load to cause deposition much beyond the wide, inner part of the shelf or that material by-passes the shelf.

The suggestion that there may be little active sedimentation on the outer part of the northwestern Gulf of Mexico continental shelf is of considerable interest. West of Galveston active sedimentation in any quantity seems to be occurring for a distance of approximately 30 mi. offshore, out to depths of 30-75 m., whereas essentially no sedimentation is indicated offshore from Mississippi Sound. Both areas have wide barrier islands, and those in the Texas area are more continuous than those along the Mississippi coast. The Texas bays would appear to provide a better trap for sediment than does Mississippi Sound, on the basis of geography alone. The possibility that this trapping of sediment in the Texas lagoons is not so effective may be related to the intense river floods characteristic of this area. Active deposition of sediment almost twice as far offshore in the Atchafalaya Bay area as off Texas may be explained by the large runoff and load of the Atchafalaya and Mississippi rivers.

Slow sedimentation on the outer shelf does not appear to be confined to the northern Gulf of Mexico, but information is scanty. Slow rates of deposition are shown off the west coast of Baja California by Walton (1955). Another probable example of little or no outer shelf deposition is given in Houbolt's (1957) report on the surface sediments from the Qatar Peninsula area in the Persian Gulf. Photograph 23 of his *Rotalia—Elphidiella* assemblage clearly shows an innermost continental shelf assemblage which Houbolt reports as characteristic of the center of the Persian Gulf at depths in excess of approximately 35 fm. (64 m.). It appears that deposition has not covered this material which was deposited during a previous cycle when sea-level was considerably lower. M. Poornachandra Rao (personal communication) has convincing evidence of little or no present-day sedimentation on the outer shelf in a part of the Bay of Bengal. His outer shelf faunas also contain abundant inner shelf forms mixed with outer shelf species, and these are in the surface sediment. Slow deposition may be characteristic of the outer shelf in many areas of the world except off deltas. Possible explanations for non-deposition on the outer shelf are: (1) either the materials now being supplied do not get that far, perhaps because of low supply, being trapped behind barriers, or because of rapid settling; or (2) the sediments are by-passing the outer shelf because of some process there which prevents their deposition. Long-period reflected waves, called "surf-beat" by Munk (1949), may aid in preventing significant deposition on the outer shelf.

The surface of apparent non-deposition on the outer shelf in many places may be late Pleistocene or early Recent in age. If so, it may be thought of as a kind of potential "unconformity." While some of the organic remains represent the modern environment, some appear to represent a previous one, and little of the terrestrially-derived sediment may have been emplaced there under present conditions.

One striking feature of the sediments in the presumed areas of "non-deposition" described above is the generally high total populations of Foraminifera in the majority of the samples. Such high populations seem to represent accumulation of tests over a long period of time with relatively less dilution from inorganically-derived sediment than in areas of more rapid deposition. High populations of dead Foraminifera probably are indicative of areas where little deposition of detrital sediment has occurred in the marine environment for a relatively long time and may not indicate any particular environment.

An interesting possible example of an area of residual sediment occurs off San Diego, California, on Coronado Bank. The Foraminifera from this area have been analyzed by Butcher (1951). The sediments on the top of Coronado Bank, at depths less than 100 fm. (183 m.), are composed largely of Foraminifera. The populations in this shoal area average more than 100,000 per 10 ml. of sediment, whereas the populations in Loma Sea Valley on the east and on the flanks of San Diego Trough on the west are much smaller (see Figure 62).

Emery *et al.* (1952) have noted that the sediments of Coronado Bank contain appreciable quantities of authigenic minerals such as glauconite, phosphorite, and collophane. They state (*op. cit.*, p. 543): "The detrital constituents of the Coronado Bank sediments are considerably more rounded and more highly weathered than those of the inshore provinces. It is inconceivable that the coarse sediments could be transported across Loma Sea Valley and deposited on top of

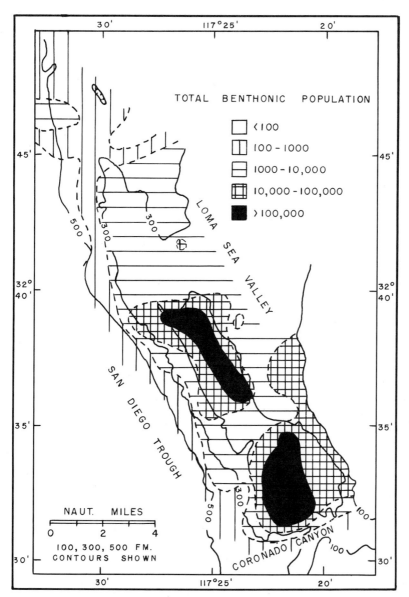

Figure 62. Total populations of benthonic Foraminifera off San Diego, California. After Butcher (1951).

Coronado Bank under present conditions. The same general suite of heavy minerals characteristic of the inshore areas is also present in the bank sediments, but there are some notable differences in their percentages. Highly altered grains are fairly common in the bank samples, but no grains of this type were found in samples from the inshore areas. Similarly, sphene, epidote, and zoisite are considerably more abundant on Coronado Bank, and magnetite and hypersthene, which are common in the inshore areas, are rare on Coronado Bank. These factors suggest that the bank sediments have a different source and are of greater age than some of the inshore sediments."

Walton (1955) has indicated areas of residual faunas in the Todos Santos Bay area, Baja California. These are characterized by very large dead populations, very low living-total population ratios, and the abundance of forms which are not now living at the depth at which the residual fauna was collected. The enclosing sediment is coarse and it is tentatively interpreted as a turbulent-zone deposit originally laid down when sea-level was 25-30 fms. (46-55 m.) lower than at present.

Seasonal Production

One of the most comprehensive analyses of living populations of Foraminifera is by Walton (1955) from Todos Santos Bay, Baja California. The total number of living Foraminifera in that area reaches a maximum in August and there is a secondary maximum in June (Figure 63). These maxima are believed to be related to the late spring and late summer flowering of diatoms. His average populations in the open ocean plotted against depth vary from approximately 20 to more than 200 living specimens per sample (Figure 67), indicating that different environments may have different standing crops and presumably different production rates.

Reiter's (1959) seasonal study of intertidal populations in Santa Monica Bay, California, shows the largest living population in fall and a decrease in the winter months.

F. L. Parker and Athearn (1959) found the largest standing crops in the marshes of Poponesset Bay, Massachusetts, during June and the lowest in December. The forms which

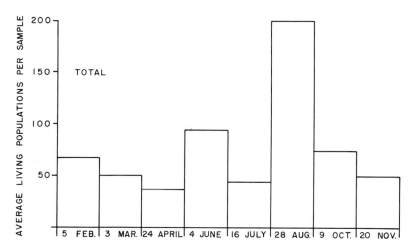

Figure 63. Seasonal abundance of average living populations in Todos Santos Bay, Baja California, Mexico. After Walton (1955).

showed a marked increase in abundance seasonally, suggesting that reproduction may occur seasonally, are *Ammotium salsum* (Cushman and Brönnimann) in August, *Elphidium galvestonense* Kornfeld in June and September, and *Protelphidium tisburyense* (Butcher) in June. Other species in this marsh area maintain a fairly constant standing crop, suggesting that they reproduce more frequently, although there is a general decrease in most of them during winter months.

Seasonal collections for living Foraminifera were made at 32 stations in Aransas, Mesquite, and San Antonio bays on the central coast of Texas (Phleger and Lankford, 1957). These were made during August and November, 1954, January, March, May, and June, 1955. There is no uniform relationship between size of living population and season of collection at the lower bay stations. Each of these stations seems to have its season of largest population more or less independent of nearby stations and no generalizations can be made about the time of largest standing crop in this part of the area. In upper San Antonio Bay the average populations for November and January are almost twice as large as those for other seasons in the upper bay. The explanation for this apparent seasonal abundance is not clear.

Size-distribution histograms of the individuals in a population were made at many stations for *Streblus beccarii* (Linné) and *Ammotium salsum* (Cushman and Brönnimann) (*op. cit.*), since these species have the highest consistent frequencies. Size distributions of these two species (Figures 64-65) show that there is generally a large range in size at all stations at all seasons, and moreover, that the spread in distribution is similar at all seasons at the same station. This suggests that the populations of these species are composed of mixed age groups. It appears, therefore, that reproduction is occurring at frequent intervals and at all seasons of the year. This reproduction habit may be characteristic of many species in the lagoon environment.

Frequent reproduction is in agreement with results obtained by Bradshaw (1957) on laboratory cultures of *Streblus beccarii*. He finds that under laboratory conditions this species will reproduce every 4 to 8 weeks within the ranges of temperature and salinity to be expected in the environment represented by the Texas lagoons.

In contrast to this, some species of *Quinqueloculina* show evidence of seasonal reproduction. This is best shown by the

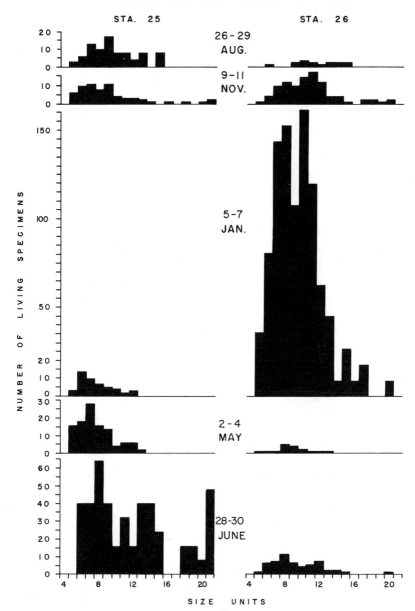

Figure 64. Size distribution histograms of living *Ammobaculites salsus* Cushman and Bronnimann and vars. [= *Ammotium salsum*] at stations 25 and 26 in San Antonio Bay area, Texas. After Phleger and Lankford (1957). Size units are 0.025 mm.

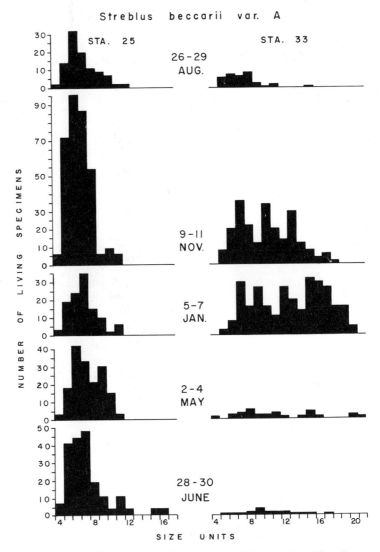

Figure 65. Size distribution histograms of living *Streblus beccarii* (Linné) var. A at stations 25 and 26 in San Antonio Bay area, Texas. After Phleger and Lankford (1957). Size units are 0.025 mm.

size distribution of *Q. poeyana* d'Orbigny at sta. 3 (Figure 66) where there was a large population of relatively small specimens in August, 1955, and by March of the following year the total population had decreased and the average size of the individuals had increased. In the following June the average size of individuals was again relatively smaller. This indicates that summer is the season of reproduction for this species at sta. 3.

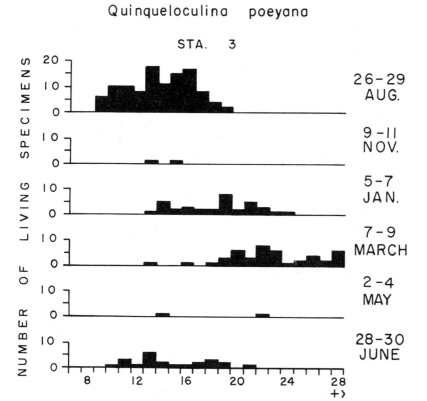

Figure 66. Size distribution histograms of living *Quinqueloculina poeyana* d'Orbigny at station 3 in San Antonio Bay area, Texas. After Phleger and Lankford (1957). Size units are 0.025 mm.

HIGH PRODUCTION AREAS

The size of the standing crop of living Foraminifera may be an indicator of some marine processes. Unusually large standing crops are of particular interest. Most of the samples which have been examined at the Marine Foraminifera Laboratory have a living population of approximately 50-200 specimens/standard sample (10 ml. wet sediment). Data from the Todos Santos Bay area in Mexico, for example (Figure 67), show an average population size of 150-200/sample. The average population size in San Antonio Bay, Texas, of 100-200 specimens/sample may be characteristic of many coastal lagoons. In certain areas, on the other hand, populations of more than 1000/sample are common. Living populations intermediate between these values are relatively infrequent and have no consistent distribution.

In the deltaic marine environment of the Mississippi Delta, populations are all more than 1000/sample and range up to 3500/sample. Organic productivity in this area is extremely high (Thomas and Simmons, 1960), with a fixation of more than 200 ugC/L/day. Populations as large as 2600/sample were found off the Guadalupe River in San Antonio Bay, Texas, in a similar environment.

Surprisingly large populations of living Foraminifera have been found in the two hypersaline lagoons which have been studied. In Laguna Madre, Texas, populations are as high as 1640/sample, and are much higher than in neighboring Aransas Bay, Mesquite Bay and most of San Antonio Bay. Certain parts of Laguna Madre support an unusually large standing crop of molluscs and fish, and the lagoon appears to be an area of very high organic production. Laguna Ojo de Liebre in Baja California, Mexico, also is a hypersaline lagoon, and some populations of living Foraminifera are as high as 1650/sample. Measurements of carbon fixation have shown that this lagoon is very productive.

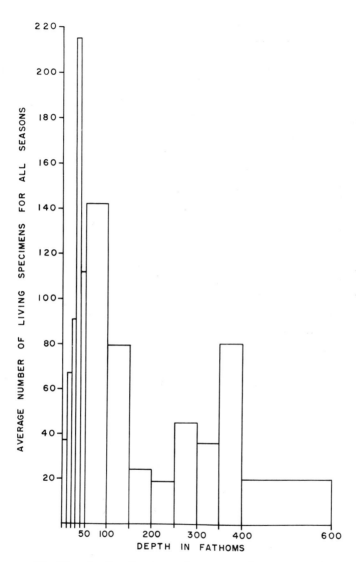

Figure 67. Depth distribution of average living populations in Todos Santos Bay, Baja California, Mexico. After Walton (1955).

Off San Diego, California, there is a large area containing more than 1000 specimens/sample in a zone parallel to the coast between 55 and 100 fm. (100-183 m.) water depth (Uchio, 1960). The large standing crop may be related to upwelling of nutrient-rich water in this area and consequent high production. The largest standing crops in the nearshore, shallow-water zone, less than 20 fm. (55 m.), are found in areas of abundant vegetation. Very large living populations are recorded from some places in marine marshes, also an environment with abundant vegetation.

Especially large standing crops, therefore, are seen off some active deltas, in hypersaline lagoons, in upwelling areas and in thickly vegetated areas. These are environments having high rates of organic production. It is suggested that the principal factor governing the size of the standing crop of living Foraminifera is quantity of available food. This is in agreement with some experimental results of Bradshaw (1955) who showed that an increase in food caused a marked increase in population growth. The environments listed above contain abundant phytoplankton and zooplankton.

CHAPTER V

Studies of Planktonic Foraminifera

In the study of a sample of Foraminifera the planktonic and the benthonic populations should be considered separately. These two assemblages represent different habitats and their occurrence together in marine sediments is, in a sense, a fortuitous association. There are fewer distinct forms of planktonic than of benthonic Foraminifera, probably totalling not more than 30-50 modern species. Individual species are widely distributed by ocean currents, and occur in marine water masses which are suitable to their survival and propagation. Most of the recognized modern species show great variation and it is possible that some variants are in reality distinct and should be given specific or subspecific rank. Planktonic Foraminifera are important contributors to marine sediments and may constitute a large per cent of sediment where rate of deposition of inorganic sediment is slow, such as in the deep sea; and thus a knowledge of their ecology is essential to interpretation of such sediments.

Most studies of distributions of modern planktonic Foraminifera have been based on their occurrence in surface sediment samples. Such studies may be valid only in a general way. It is more reasonable to analyze their distribution in the water masses in which they live. If empty planktonic tests are placed in a tube of sea water they settle rapidly, and on this experimental evidence it may be assumed that their occurrence in bottom sediments gives a true picture of their actual distribution in the overlying water. Such an experiment does not necessarily indicate the settling rate of these shells in the ocean. Under actual marine conditions settling can be retarded by turbulence, by current action, and by surface divergence of the water. While these processes are most active near the sea surface, they also operate to a lesser degree at greater depths. It is quite possible that the test of a planktonic foraminifer may be kept afloat and transported a considerable distance out of the environment in which it lived.

Distribution of Living Planktonic Foraminifera

Bradshaw (1959) has studied the distribution of Foraminifera in more than 700 plankton tows from the north and equatorial Pacific. Four distinct faunas can be distinguished which have discrete distributions, a subarctic fauna, a transition fauna, a central fauna, and an equatorial west central fauna. The composition of these faunas is listed on Figure 68 and their geographic distribution is shown on Figure 69. These faunas appear to be characteristic of surface water masses which also can be distinguished by temperature and salinity characteristics. The largest populations occur in subarctic areas and at some places in the equatorial region, with the central fauna having the smallest populations. The regions of abundant Foraminifera are also areas of high phosphate concentration. Specimens are most abundant in the upper 100 m. of water. F. L. Parker (1960) has found similar faunal distributions in 81 plankton tows from the equatorial and south-

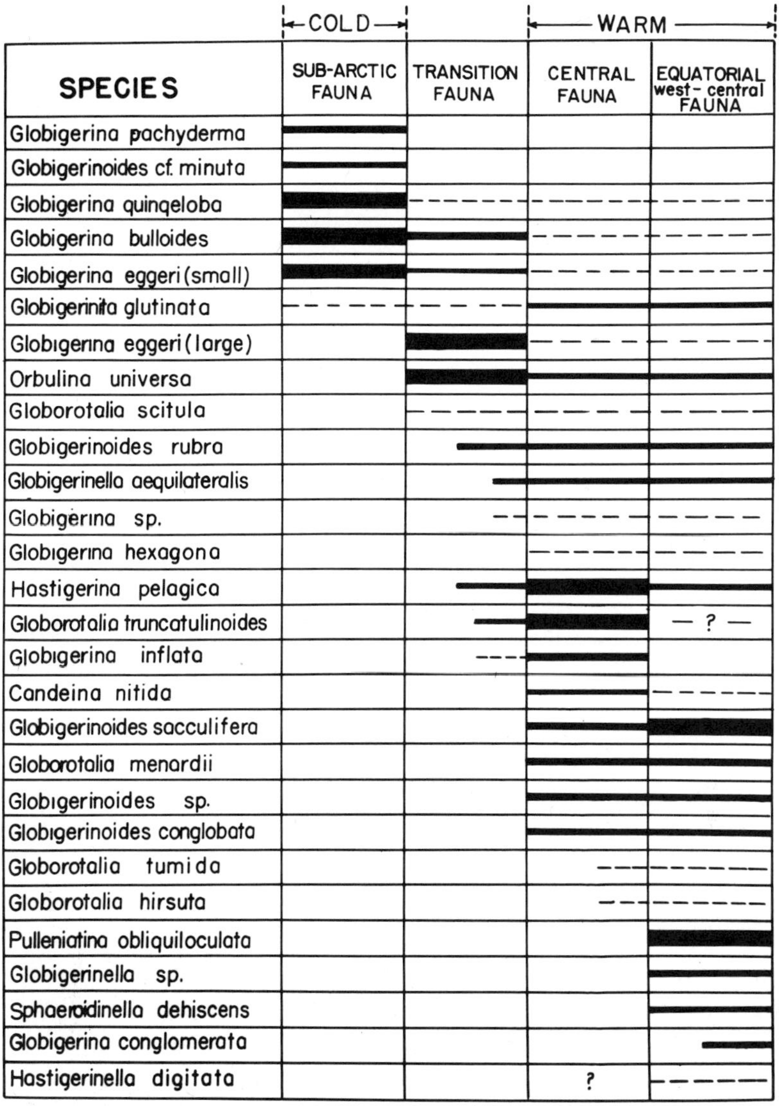

CHAPTER V   215

SPECIES	SUB-ARCTIC FAUNA	TRANSITION FAUNA	CENTRAL FAUNA	EQUATORIAL west-central FAUNA
Globigerina pachyderma	■			
Globigerinoides cf. minuta	■			
Globigerina quinqeloba	■	- - -	- - -	- - -
Globigerina bulloides	■	■	- - -	- - -
Globigerina eggeri (small)	■	■	- - -	- - -
Globigerinita glutinata	- - -	- - -	■	
Globigerina eggeri (large)		■	- - -	- - -
Orbulina universa		■	■	
Globorotalia scitula		- - -	- - -	- - -
Globigerinoides rubra		■	■	
Globigerinella aequilateralis		■	■	
Globigerina sp.		- -	- - -	- - -
Globigerina hexagona			- - -	- - -
Hastigerina pelagica		■	■	
Globorotalia truncatulinoides			■	— ? —
Globigerina inflata		- - -	■	
Candeina nitida			■	- - -
Globigerinoides sacculifera			■	■
Globorotalia menardii			■	■
Globigerinoides sp.			■	■
Globigerinoides conglobata			■	■
Globorotalia tumida			- - -	- - -
Globorotalia hirsuta			- - -	- - -
Pulleniatina obliquiloculata				■
Globigerinella sp.				■
Sphaeroidinella dehiscens				■
Globigerina conglomerata				■
Hastigerinella digitata			?	- - -

Figure 68. Composition of North Pacific planktonic foraminiferal faunas based on plankton tows. After Bradshaw (1959).

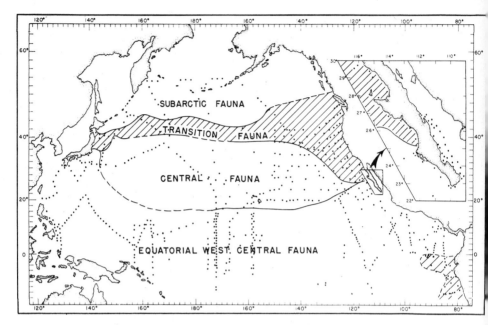

Figure 69. Generalized distribution of North Pacific planktonic foraminiferal faunas based on plankton tows. After Bradshaw (1959).

east Pacific. The use of planktonic Foraminifera to trace the distribution of oceanic water masses is amply illustrated by these Pacific studies.

Preliminary studies of distribution of planktonic Foraminifera in the Atlantic area have been made by Schott (1935), Phleger (1945, 1951b) and Bé (1959). Living planktonic Foraminifera have been obtained from all depths sampled in the Atlantic down to at least 2000 m. The largest population, however, was obtained from the upper 100 m. by Schott (1935) (average of 489 specimens per tow). In the northwestern Gulf of Mexico the living population in the upper layer of water was 5-6 specimens/cu. m. of water at most stations. Three tows from the Sigsbee Deep have the following number of specimens/cu. m.: 3/cu. m. at 43 m. from Station 341, 2/cu. m. at 40 m. from Station 142, and 73/cu. m. at 40 m. from Station 148. Station 148 is only about 25 miles from Station 142, and the depth of water and temperature and salinity are essentially the same. Such "patchiness" in population density and composition is well-known to marine planktonologists.

At a few stations a larger living population has been recorded from a considerable depth than from the upper water layers. This is best illustrated by the following from the northwestern Gulf of Mexico (Phleger, 1951b, p. 35):

Station	Depth (m.)	Living Foraminifera
43	30	6.5/cu. m. water
43	340	305/cu. m. water
451	35	1.6/cu. m. water
451	425	17.5/cu. m. water

At other stations in the area the populations are essentially the same at all depths sampled. Occasional specimens of planktonic forms containing protoplasm have been collected in bottom sediment samples; this suggests that some specimens may live a part of their lives on or near the bottom.

Most plankton samples contain a few empty tests of Foraminifera. At many stations there are more empty tests in the deeper samples, and at some stations the opposite is true. The presence of significant populations of these empty tests in the upper layers suggests that some empty planktonic tests do not sink immediately.

These distributions have an important bearing on the interpretation of planktonic assemblages occurring in bottom sediments. It is not necessarily correct to assume that the entire assemblage was adjusted to the environment in the upper layer of water. The fauna in the sediment may have been adjusted to a series of different environments existing throughout the entire column of water. It is possible, nevertheless, that an accumulated assemblage may reflect near-surface conditions, at least in large part.

Emiliani (1954) has studied the oxygen isotope composition of a few species of planktonic Foraminifera from sediment samples to determine the water depth at which they lived. This method determines the water temperature at which the calcium carbonate formed the foraminiferal tests. The depth of water at which this occurred is estimated on the basis of an average relationship between depth and temperature in the region from which the material came. His conclusions are as follows (*op. cit.*, p. 158):

"The zones of maximum densities of populations of pelagic Foraminifera are stratified in a definite order with respect to temperature and water density, and therefore also, indirectly, with respect to depth. Different populations of the same species show small tolerance toward variations of temperature and water density, and may occupy considerably different depths in different regions, in order to adjust themselves to the more important temperature and water density factors.

"In the material so far examined, no evidence has been found of populations living at depths greater than about 220 m.

"All species examined, except *Orbulina universa*, maintain unchanged their depth habitats during at least most of their lives. On the other hand, *Orbulina universa* occupies progressively shallower depth habitats while growing.

"Finally, the considerable reluctance of foraminiferal populations to vertical displacements, in a given area, and the independence of these displacements from at least small temperature variations, enhance the possibility of using pelagic Foraminifera from cores for paleoclimatological studies."

Some of Emiliani's generalizations do not agree with what is known of the distribution of planktonic Foraminifera from collections in plankton tows. These forms have been found living at depths greater than 220 m., and in some places relatively large populations have been collected from much greater depths. The isotope method reflects the temperature at which the calcium carbonate of the test was precipitated. It is possible that the specimens living in deep, cold water were precipitating little or no shell material at those depths and had gained most of their tests in the upper water layers. The depth of habitat (temperature) indicated by Emiliani's results may represent only an average of several depths at which specimens collected in a sediment sample lived. If this is true, it certainly indicates that the highest population lives in the upper layers of water.

The interpretation of the depth habitat based upon oxygen isotope analyses depends in addition upon accurate knowledge of bathymetric distribution of temperature in the area where the animals lived. Knowledge of bathymetric distribution of ocean temperatures is inadequate in some areas. The temperature depth curves given by Emiliani (*op. cit.*, p. 151) for the eastern Gulf of Mexico and the tropical Atlantic may not be the best generalizations possible for these areas. The symbiotic algae known to occur in many Foraminifera may cause oxygen exchange with the calcium carbonate of the test. If this occurs, as seems likely, oxygen isotope composition may be somewhat modified by purely biological processes after its deposition in the test.

Le Calvez (1936) has made some interesting observations on the vertical distribution of *Orbulina universa* d'Orbigny in the Mediterranean. He found "normal" specimens near the surface containing the internal globigerinid test centered in the spherical external chamber. At 50 m. the internal shell is depressed against the side of the test and the shell appears to be thinner, and at 100 m. the internal test is not present. The disappearance of the internal, globigerinid test appears to be related to reproduction of this form. The observations indicate that *Orbulina* migrates into deeper water at some time preceding reproduction and much of the actual reproduction may occur at approximately 300 m. The new generation of *Orbulina* produced in deep water probably migrates to the upper layers by the time the adult stage is reached. Emiliani's results (1954, p. 154) tend to confirm this, in that his larger size group of *Orbulina* gave a temperature of 1.5-3°C. higher than the smaller size group. Some of the calcium carbonate in the shell of *Orbulina* is formed through a range of depths and temperatures, and the temperatures indicated by the oxygen isotope method probably are averages of these. This is suggested, for example, in Emiliani's temperature differences (1.5-3°C.) in his size groups. This difference is much too small to account for 250-300 m. of vertical migration as Le Calvez' work indicates.

Distributions in the North Atlantic Based on Sediment Samples

As stated previously, most of our knowledge of the distribution of planktonic Foraminifera is based on their occurrences in the surface sediments. Schott (1935) has given excellent quantitative data on the distributions of species in samples collected by the *Deutsche Atlantische Expedition*. Schott's stations are between Africa and South America and extend from approximately 8° S. Lat. to 18° N. Lat. These are typical low-latitude faunas, dominated by high frequencies of *Globigerinoides sacculifer* (H. B. Brady), *Globorotalia menardii* (d'Orbigny), *G. tumida* (H. B. Brady), *Globigerina eggeri* Rhumbler and *Pulleniatina obliquiloculata* (Parker and Jones).

Ovey (*in* Wiseman and Ovey, 1950, p. 59, Table 1) studied nine samples from the Atlantic ranging from 66° S. Lat. to 69° N. Lat. and lists the distribution of the species in terms of their "temperature distribution":

Arctic and Antarctic

> *Globigerina dutertrei* d'Orbigny
> *G. pachyderma* (Ehrenberg)

Temperate

> *Globigerina bulloides* d'Orbigny
> *G. inflata* d'Orbigny
> *Globorotalia crassula* Cushman and R. E. Stewart
> *G. canariensis* (d'Orbigny)
> *G. hirsuta* (d'Orbigny)
> *G. truncatulinoides* (d'Orbigny)

Warm

> *Orbulina universa* d'Orbigny
> *Globigerina dubia* Egger
> *Globigerinella aequilateralis* (Brady)
> *Globigerinoides ruber* (d'Orbigny)
> *G. sacculifer* (Brady)
> *G. conglobatus* (Brady)
> *Globorotalia menardii* (d'Orbigny)
> *G. tumida* (Brady)
> *G. scitula* (Brady)
> *Sphaeroidinella dehiscens* (Parker and Jones)
> *Pulleniatina obliquiloculata* (Parker and Jones)

Schott (1952) has listed the following as belonging to the Atlantic warm-water fauna:

> *Globigerinoides sacculifer* (Brady)
> *G. ruber* (d'Orbigny)
> *Globigerina dubia* Egger
> *Globigerinella aequilateralis* (Brady)
> *Orbulina universa* d'Orbigny
> *Globorotalia menardii* (d'Orbigny)
> *G. tumida* (Brady)
> *G. truncatulinoides* (d'Orbigny)

His cold-water fauna is composed principally of *Globigerina bulloides* d'Orbigny and *G. inflata* d'Orbigny.

Phleger *et al.* (1953) studied the distribution of planktonic Foraminifera in fifty-three surface sediment samples from approximately 0° Lat. to 50° N. Lat. in the North Atlantic. They recognized five generalized distributions (Plate 10).

Species occurring in high frequencies in latitudes lower than 20° N. Lat., and also present in low frequencies at a few mid-latitude stations:

>   *Globorotalia menardii* (d'Orbigny)
>   *G. tumida* (Brady)
>   *Pulleniatina obliquiloculata* (Parker and Jones)

Species more abundant in low latitudes than mid-latitudes but present at most stations:

>   *Globigerina eggeri* Rhumbler
>   *Globigerinoides sacculifer* (Brady)

Globigerina pachyderma is confined to high and mid-latitudes.

Species more abundant in mid-latitudes but also present in low latitudes:

>   *Globigerina bulloides* d'Orbigny
>   *G. inflata* d'Orbigny
>   *Globorotalia hirsuta* (d'Orbigny)
>   *G. scitula* (Brady)
>   *G. truncatulinoides* (d'Orbigny)

Species universally distributed:

>   *Globigerinella aequilateralis* (Brady)
>   *Globigerinita glutinata* (Egger)
>   *Globigerinoides conglobatus* (Brady)
>   *G. ruber* (d'Orbigny)
>   *Orbulina universa* d'Orbigny

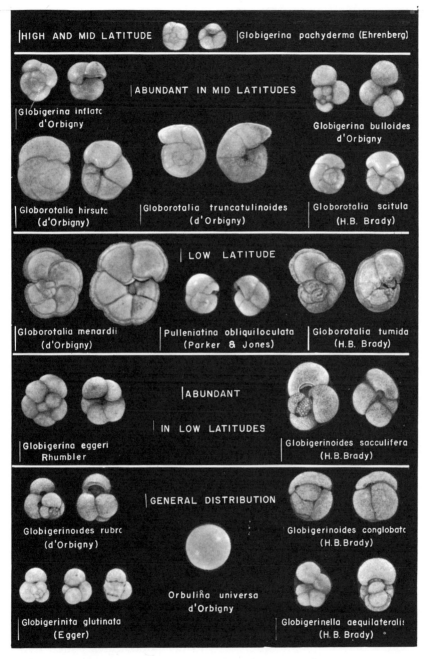

Plate 10. Species of common North Atlantic planktonic Foraminifera arranged according to types of distribution. After Phleger (1954b).

There is an approximate correlation between distribution of surface isotherms and latitude. The distribution of the mean sea surface temperatures in the North Atlantic (Figure 1) for August and February is based upon the sea-surface temperature charts compiled by the U. S. Hydrographic Office. It should be emphasized that while these charts are statistically correct they are broad generalizations and should be evaluated as such. They do not show actual and extreme variations in surface temperatures which undoubtedly are of primary importance in regulating the distribution of organisms such as the planktonic Foraminifera. Moreover, they do not give information on the vertical distribution of temperature which may be of importance in affecting distribution of planktonic Foraminifera. The correlation between the distributions of species of planktonic Foraminifera and the mean monthly surface isotherms is not good in a statistical sense. It is sufficiently indicative, however, to suggest that the surface temperatures are of importance in the ecology of these species.

The near-surface water is expected to have the greatest ecological effect on the populations because these are planktonic organisms which seem to have their largest population in the near-surface layers of marine water (upper 100 m.). Surface temperature is an obvious and important feature of near-surface oceanic water masses. Temperature is useful as at least a tentative indicator of such water, and certainly does have direct and indirect effects on the populations inhabiting the upper 100 m. The possible effects of the deeper waters on planktonic populations which inhabit them cannot be evaluated at the present time. Plankton tow data suggest that any population accumulating in the sediment may record environments of the entire water column. Since these deeper populations appear to be much smaller than the near-surface populations, the ecologic effects of deeper water may be largely masked by the more abundant forms from the upper layers. There may be slight differences in the populations which live in the deeper layers but no information is available in this connection.

Mixed Planktonic Faunas

Mixing of planktonic faunas occurs wherever there is mixing of the different types of water which these faunas inhabit. One of the most obvious examples of this is the transport of low-latitude species, such as *Globorotalia menardii* (d'Orbigny), into mid-latitudes by the Gulf Stream system in the North Atlantic. The transport of warm water into mid- and high-latitudes by this mechanism is well-known. Occasional masses detach themselves from the Gulf Stream and invade far beyond what appears to be the normal path of the currents. The presence of occasional specimens of low-latitude species, such as *G. menardii* and *Globigerinoides sacculifer* (Brady), in the southern Gulf of Maine undoubtedly is the result of such a mechanism. Occurrence of similar low-latitude species in Pleistocene sediments under the city of Boston suggests that this mechanism has been occurring for some time (Phleger, 1949; Stetson and Parker, 1942).

The Gulf Stream is bounded on the north by sub-Arctic water. The position of this boundary varies considerably and there is mixing of the Gulf Stream and sub-Arctic water in the boundary area. The sediments underlying this region contain a mixed high and mid-latitude fauna with some low-latitude species in low frequencies. A sample reported (Phleger and Hamilton, 1946, Plate 1) from this boundary region near Georges Bank contains the following species in relative abundance:

Globigerina eggeri Rhumbler (low-latitude)
G. inflata d'Orbigny (mid-latitude)
G. pachyderma (Ehrenberg) (high-latitude)
Globigerinoides ruber (d'Orbigny) (low and mid-latitude)
Globorotalia menardii (d'Orbigny) (low-latitude)

Other samples in the collections of the Marine Foraminifera Laboratory from this general area have a similar faunal com-

226   *Ecology and Distribution of Recent Foraminifera*

position. Such a natural mixture of these faunal elements could only occur at the position of the Arctic or Antarctic convergence; low-latitude species are expected to be more abundant on the upstream side of the warm-water current in such a convergence area.

There is a high concentration of *Globigerina bulloides* d'Orbigny, a typical mid-latitude species, at a few stations off the west coast of Africa between Cape Verde and Cape

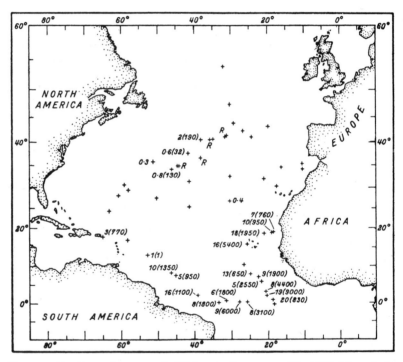

Figure 70. Distribution of *Globorotalia menardii* (d'Orbigny) in surface sediment samples from the North Atlantic. First number at each station is per cent of total population of planktonic Foraminifera; second number, in parentheses, is estimated number of specimens. After Phleger, Parker and Peirson (1953).

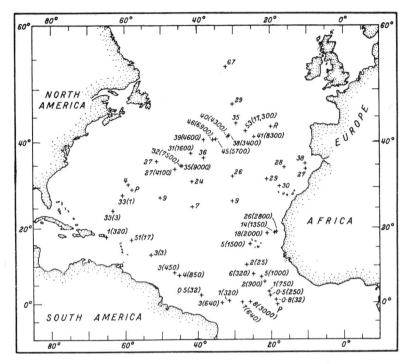

Figure 71. Distribution of *Globigerina bulloides* d'Orbigny in surface sediment samples from the North Atlantic. First number at each station is per cent of total population of planktonic Foraminifera; second number, in parentheses, is estimated number of specimens. After Phleger, Parker and Peirson (1953).

Blanco, and a high frequency of *Globorotalia menardii* (d'Orbigny), a low-latitude form, at the same stations (Figures 70, 71). The occurrence of these two species together in high frequencies is an apparently anomalous association. These mixed assemblages occur in the area of the Equatorial Counter Current where it combines with water from the Canaries Current to form the North Equatorial Current. This also is the position of the Northeast Trade Winds which provide most of the force moving the North Equatorial Current. These mixed faunas occur because the low-latitude forms are being contributed from the Equatorial Counter Current water and the mid-latitude forms are from the mid-latitude water of the Canaries Current. Mixed faunas in these latitudes are expected to occur only on the eastern sides of oceans in a convergence area. There is a water divergence on the opposite side of the North Atlantic at this latitude and a pure low-latitude fauna is found. The general distribution of faunas of planktonic Foraminifera in the North Atlantic is shown on Figure 72.

Distributions of several species of planktonic Foraminifera in the northern Gulf of Mexico are shown on Figures 73-79.

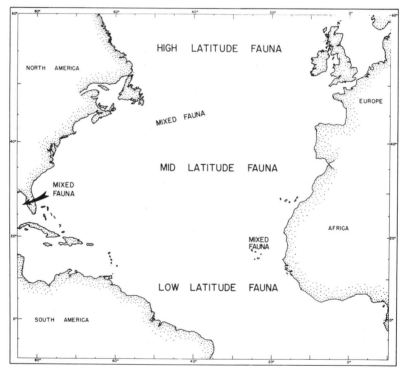

Figure 72. Generalized faunas of planktonic Foraminifera in the North Atlantic. After Phleger (1954b).

These distributions are based on occurrences in more than 700 surface sediment samples from this region and are believed to give an accurate general picture of occurrences. It is seen that the Gulf of Mexico fauna is a mixed low-latitude and mid-latitude one. An illustration of the mixed nature of this fauna is the association of the following species in relatively high frequencies: *Globigerina bulloides* d'Orbigny (Figure 73), *G. eggeri* Rhumbler (Figure 74), *Globigerinoides sacculifer* (Brady) (Figure 75), *Globorotalia menardii* (d'Orbigny) Figure 76), *G. truncatulinoides* (d'Orbigny) (Figure 77) and *Pulleniatina obliquiloculata* (Parker and Jones) (Figure 78). Most or all of the water being supplied to the Gulf of Mexico appears to be from the North Equatorial Current system and should contain a pure low-latitude fauna. It is expected that the Gulf of Mexico would be populated with this assemblage. Two possible explanations are suggested for the presence of the mixed fauna: (1) Mid-latitude water carrying its indigenous planktonic fauna may enter the Gulf of Mexico around the Florida Peninsula. It is possible that this water enters and is dispersed by a coastal current system flowing west and northwest. (2) The mid-latitude species may have invaded this region during one of the Pleistocene glacial stages, or at some other time in the past, and may now be a relict fauna.

There appear to be two types of distribution in the northern Gulf of Mexico. *Globigerina eggeri* Rhumbler (Figure 74), *Globigerinoides sacculifer* (Brady) (Figure 75) and *Globorotalia menardii* (d'Orbigny) (Figure 76) generally occur in highest frequencies in the southeast part of the area, opposite Yucatan Strait where the tropical water enters the Gulf of Mexico. The frequencies of these species decrease toward the north and northwest. This suggests that the Gulf of Mexico is being stocked with these three species as a result of their being introduced with the low-latitude water which they inhabit. These low-latitude species would probably be of uniformly high frequency in the region if low-latitude water filled the entire basin.

The characteristic mid-latitude species, *Globigerina bulloides* d'Orbigny (Figure 73) and *Globorotalia truncatulinoides* (d'Orbigny) (Figure 77), are more abundant in the northwesternmost Gulf of Mexico and decrease in frequency toward the east and south. The center of highest frequency does not appear to be related to any immediate source of mid-latitude water from the Atlantic. It is possible that these species were introduced into the Gulf of Mexico in abundance during some earlier time when abundant mid-latitude water was supplied to the basin. They may have been able to survive and reproduce because the northwest Gulf of Mexico assumes characteristics of mid-latitude water during the winter season. The distribution of the low-latitude and mid-latitude species suggests that the water in the northwest behaves as a separate mass. This is in agreement with a suggestion made by R. C. Reid (personal communication) that the water circulation in the western Gulf of Mexico is in the form of a large eddy and that there may be little communication with the water in an eddy occupying the eastern part of the basin. It is apparent from the distributions of the planktonic Foraminifera that some water interchange occurs but it may be relatively small.

The existence of two general "cells" of offshore water in the northern Gulf of Mexico is further suggested by examination of the current charts published by the U. S. Navy Hydrographic Office (No. 225, 1947). According to these charts, much of the low-latitude water which enters the Strait of Yucatan flows directly north and east and leaves the Gulf of Mexico through Florida Strait. Some of the low-latitude water appears to diverge south of the Mississippi Delta, with part of the water flowing west and more or most of it flowing to the east. The eastward-flowing water turns southward and joins the water from the Yucatan Strait, flowing through the Florida Strait to form the Florida Current. The water in the western Gulf of Mexico appears to act as a more or less separate circulation system with only a small amount of the water originat-

(Text continued on page 238.)

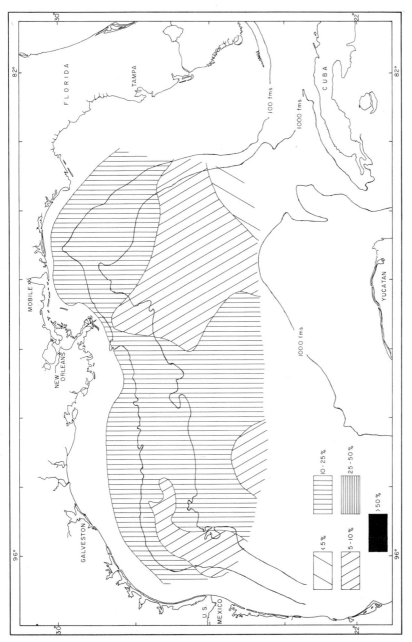

Figure 73. Distribution of *Globigerina bulloides* d'Orbigny in northern Gulf of Mexico in per cent of total population of planktonic Foraminifera. After Phleger (1954b).

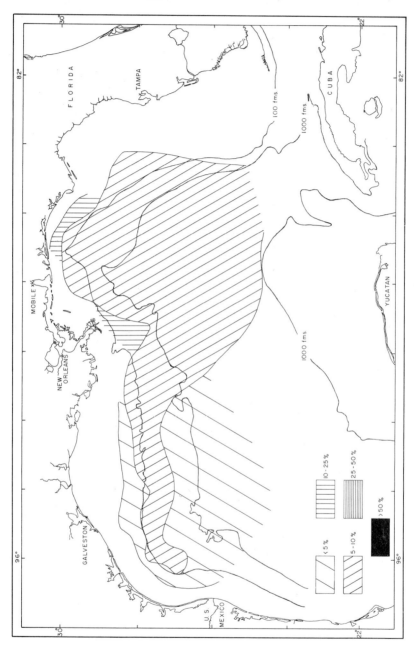

Figure 74. Distribution of *Globigerina eggeri* Rhumbler in northern Gulf of Mexico in per cent of total population of planktonic Foraminifera. After Phleger (1954b).

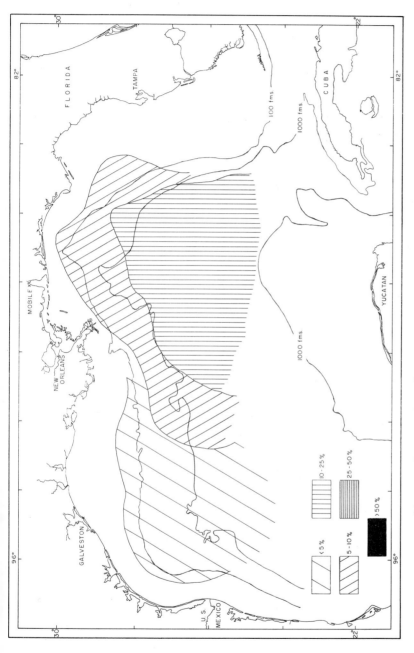

Figure 75. Distribution of *Globigerinoides sacculifer* (H. B. Brady) in northern Gulf of Mexico in per cent of total population of planktonic Foraminifera. After Phleger (1954b).

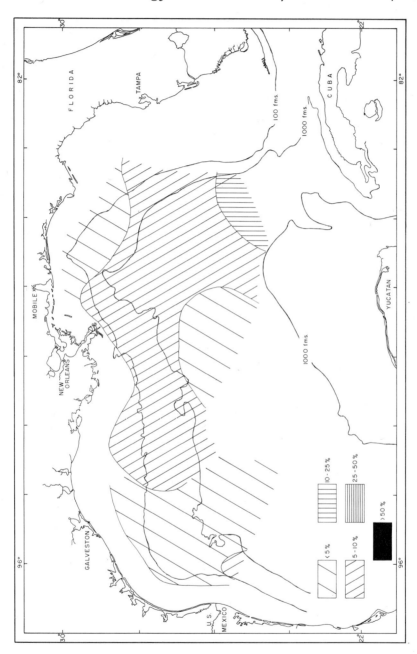

Figure 76. Distribution of *Globorotalia menardii* (d'Orbigny) in northern Gulf of Mexico in per cent of total population of planktonic Foraminifera. After Phleger (1954b).

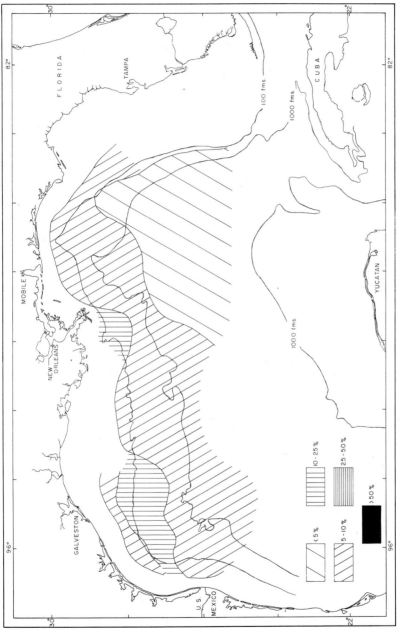

Figure 77. Distribution of *Globorotalia truncatulinoides* (d'Orbigny) in northern Gulf of Mexico in per cent of total population of planktonic Foraminifera. After Phleger (1954b).

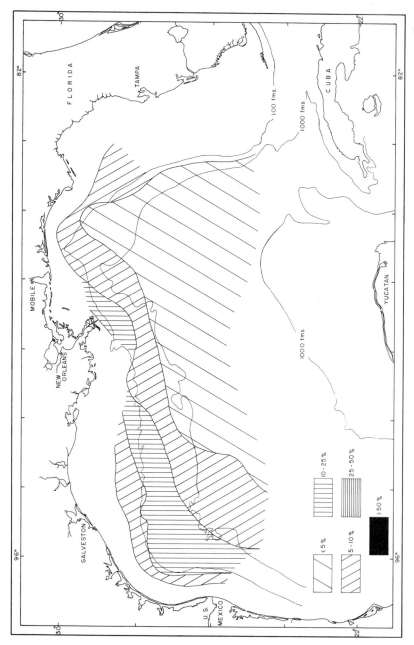

Figure 78. Distribution of *Pulleniatina obliquiloculata* (Parker and Jones) in northern Gulf of Mexico in per cent of total population of planktonic Foraminifera. After Phleger (1954b).

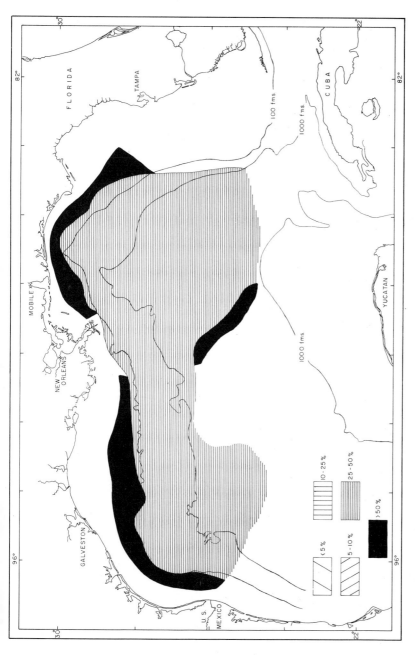

Figure 79. Distribution of *Globigerinoides ruber* (d'Orbigny) in northern Gulf of Mexico in per cent of total population of planktonic Foraminifera. After Phleger (1954b).

ing from the divergence of the water coming in from low latitudes during any period of time. The frequencies of the planktonic Foraminifera seem to reflect this general current system.

Pulleniatina obliquiloculata (Parker and Jones) is a low-latitude species which occurs in the area immediately north of Yucatan Strait (Figure 78) in frequencies similar to those in the North Atlantic. It is more abundant, however, in the northwest Gulf of Mexico, constituting up to 25% or more of the planktonic Foraminifera at numerous stations. This species may have become established in the northwest area during previous conditions. It is suggested that *P. obliquiloculata, Globigerina bulloides* d'Orbigny, and *Globorotalia truncatulinoides* (d'Orbigny) may be in part relict in the northwestern Gulf of Mexico.

The dominant species in the northern Gulf of Mexico is *Globigerinoides ruber* (d'Orbigny) which has a frequency of more than 25% of the planktonic Foraminifera at almost every station and more than 50% at numerous stations (Figure 79). The higher frequencies in the continental shelf area, shown by the black pattern in Figure 79, are due to a low total population of which *G. ruber* constitutes the majority of specimens. The dominance of this species is accentuated in such low populations. The frequencies of *G. ruber* in the northern Gulf of Mexico are quite comparable to those in the North Atlantic, where it is generally distributed in both low and mid-latitudes.

Other species in the northern Gulf of Mexico which are uniformly distributed and constitute less than 5% of the population are:

Globigerinella aequilateralis (Brady)
Globorotalia punctulata (d'Orbigny)
G. scitula (Brady)
G. tumida (Brady)
Orbulina universa d'Orbigny

The frequencies of these species also are comparable to their frequencies in the mid- and low latitudes of the North Atlantic.

Continental Shelf Distributions

The distributions discussed above are those in offshore, "oceanic" water. Continental shelf water generally is characterized by variability in temperature and turbidity, and high organic production; the salinity may be lower than offshore values where there is land runoff. Off many coasts the water on the inner part of the shelf may be somewhat different ecologically from that on the outer part of the shelf. The shores of many small oceanic islands, on the other hand, are bathed by typical oceanic water because there is insufficient runoff to modify the nearshore water. Off arid coasts having little or no runoff, such as northern Baja California, Mexico, the oceanic water may be little modified on the inner shelf.

Planktonic Foraminifera seem to be adjusted to the environments characteristic of offshore, oceanic water. They generally do not thrive in nearshore continental shelf waters off high-runoff areas. In such areas the sediments on the outer part of the continental shelf may contain a planktonic foraminiferal fauna characteristic of offshore areas. This fauna is characterized by smaller total planktonic populations and fewer species as the shore is approached, and no planktonic forms may occur in sediments nearest shore. A typical distribution of planktonic forms in the northwest Gulf of Mexico is shown in Figure 80, which records the occurrences in a traverse extending off the Brazos River and extends from a depth of 24 m. to the Sigsbee Deep in the center of the Gulf of Mexico. The species having the highest frequencies in the population, *Globigerina bulloides* d'Orbigny and *Globigerinoides ruber* (d'Orbigny), occur on the inner shelf where populations are small. The ones with the lowest frequencies, on the other hand, occur only offshore where populations are large.

STATION	DEPTH IN METERS	TOTAL PLANKTONIC POPULATION	Candeina nitida	Globigerina bulloides	G. eggeri	G. inflata	Globigerinella aequilateralis	Globigerinoides conglobata	G. rubra	G. sacculifera	Globorotalia menardii	G. punctulata	G. scitula	G. truncatulinoides	G. tumida	Orbulina universa	Pulleniatina obliquiloculata	Sphaeroidinella dehiscens
142	3246	1,150	.	9	6				3	3				4		7	9	8
141	3381	3,000		19	3				2	1				2	5	2	6	.7
140	3273	3,000		20	3				2	3	2.9			2	5	2	6	
139	3154	2,000	—	5					2.3	2.9				4	5	3		
138	3054	16,600		7,17	4				3.1	—	46,50	44	42	4	6	.7		
137	2815	186,600		8,7	10	1.3	4.3	—	46	11			8	3	4	3.6	.7	
135	1720	3,250		8	4,13	1	—	2.4	1	—	1			4	5	9		
133	1646	21,000		9	4	1.3	2.4	—	1	—				4	16	8	3	
132	3411	7,000		13	3			3	3	—	1	—		2	4	4		
130	2395	7,000		12	4			3	3	46,46	2	4	—	1	2	2	4	
129	1810	3,000		15,23	4			3	2	43	2	3	.5		2	2	5	1
127	1353	5,000		11	4			2.3	—	50	4	3	.2			2	2	5
126	1353	8,000		6	3				1		3	4	.4			4	3	2
125	1325	8,000		14	3	.4	.8	.8	1		2	3	5		.4		6	4
124	1417	16,400		7,16	4		3,4	1.9	48	48,45	2	2		.4	3.4	8	6,12	.3
122	628	2,300						1										
121	588	4,300	.2	4,8	1		3	2	39	33	1	3	6	4		2.7	3.4	7
120	960	400		4	2	1	1			18		6		.3	2	2.4	7.7	
119	736	1,500		12			2	1	39	16		2	2	2	4		7,17	
118	267	150		4	5,12	2			2	1				4	6,6	.3		
117	73	1,200		9						30					.5		2	2
116	62	500		10,23				1	60		1							
115	57	250			3				74			3				6	5,12	16
114	51	750		10,15			4		72		2							
113	46	650		33,10					85		2,6					2		
112	48	25							67									
111	38																	
110	35	50		17					80,83		3							
109	33	40		33,20														
108	31	50		33	4				100,67									
107	29	15							100									
106	35																	
105	33	80		10					90									
104	33																	
103	32	1							100									
102	30																	
101	29																	
100	26																	
99	24																	
98	24																	
97	22																	
96	22																	
95	24	1																

Figure 80. Occurrences of planktonic Foraminifera in per cent in samples taken across the continental shelf off the Brazos River, northwestern Gulf of Mexico. After Phleger (1954b).

Bandy (1956, p. 179) has interpreted similar distributions in his traverses across the continental shelf in the northeast Gulf of Mexico as follows: "Different upper limits of depth ranges of planktonic species demonstrate another method of depth correlation. *Globigerinoides rubra* (d'Orbigny) appeared in less than 100 feet [30 m.] of water, most of the species appeared between depths of 100 and 200 feet [30-61 m.], and *Globorotalia tumida* (Brady) appeared at about 400 feet [122 m.]. These data suggest that these species, when alive, float no higher than the minimum depth indicated."

Globigerinoides ruber is the most abundant member of the planktonic population in the Gulf of Mexico. It always constitutes 25-50% of the planktonic population and occasionally is more than 50% (Figure 80). Where populations are small, as on the inner continental shelf, it is the species most likely to occur because of its abundance. It is to be expected that *G. ruber* would be present at shoaler depths nearshore because of the decrease in overall abundance of planktonic Foraminifera as the shore is approached. *Globorotalia tumida* (Brady), on the other hand, is a rare constituent of the planktonic population (Figure 80). It is to be expected only where the total planktonic population is large, as on the outer shelf and deeper. It seems obvious that these distributions of planktonic species are not valid depth ranges but that they are reflections of species frequencies related to size of population. Depth distributions of planktonic Foraminifera reported by Polski (1959) from the Asiatic shelf may also be interpreted as caused by frequencies in the population and are not true depth distributions.

Curray (1960) has found that the sediments of the outer part of the continental shelf in the northwest Gulf of Mexico are characterized by having very large percentages of planktonic Foraminifera. This was shown previously (Phleger, 1951b; Bandy, 1956; F. L. Parker, 1954) and is in part a result of the restriction of oceanic water to the outer shelf in this region.

It is also in part due to accumulation of these skeletal remains over a long period of time with little dilution by deposition of detrital materials. Where no distinct coastal water is formed, as in the low-runoff areas of southern California and northern Baja California, there is little difference in composition or size of the planktonic population toward shore. As an example of this, abundant planktonic Foraminifera are regularly collected in plankton nets off the Scripps Institution of Oceanography pier in the nearshore turbulent zone. These are identical with faunas collected farther offshore in the San Diego area. Abundant planktonic Foraminifera do not necessarily distinguish outer shelf from inner shelf sediments.

Grimsdale and Morkhoven (1955) have analyzed the ratio between planktonic and benthonic Foraminifera in the northwestern Gulf of Mexico from published data. Their plot of all the population data from this area clearly shows a marked increase in ratio of planktonic to benthonic specimens from the inner shelf to the outer shelf. These authors have experimented with the suggestion that the planktonic-benthonic ratios may be useful in estimating depth of deposition. They concluded that there is so much variation in the values that this parameter is of limited usefulness for any precise determination. Their plots of the data (*op. cit.*, fig. 1) show, in addition to the increase in ratios on the outer shelf, a very irregular distribution of ratios from 100-1000 m. but trending toward larger values. Deeper than approximately 1200 m. the values are uniformly high, being mostly more than 90% planktonic specimens.

Occasionally small masses of water will be detached from the main offshore water and will invade nearshore areas or the oceanic water will shift inshore under favorable wind conditions. Eventually this water becomes mixed with the inner-shelf water. Such water will carry its characteristic planktonic fauna and deposit it in an inner-shelf environment where these forms may not normally live. These occurrences usually are represented only by single or a few specimens per sample, but occasionally an abundant planktonic assemblage may be present where invasion of such water is common.

Interpretation of Pre-Modern Planktonic Faunas in Submarine Cores

GENERAL PRINCIPLES

The modern planktonic foraminiferal fauna occurs at the top of a deep-sea submarine core if the surface sediment is intact. It has been widely recognized that in many cores there is a lower fauna which is unlike the uppermost, modern one. This lower-core fauna usually is like the modern fauna of a higher latitude than the location at which the core was collected. These higher-latitude, pre-modern core faunas have been interpreted as representing cooler surface water temperatures than those which now exist. This inference is a reasonable one even though there is only indirect evidence of temperature adjustments in the species involved. Such lower-core faunal associations are adjusted to water masses at present characteristic of higher latitudes, and these higher-latitude water masses generally do have colder surface temperatures than lower-latitude water masses. The colder-water faunas in cores from low latitudes are like those now living in mid-latitudes; the colder-water faunas in cores from mid-latitudes are like those now living in high latitudes. The possibility of recognizing colder-water faunas in high-latitude cores seems a little remote.

HIGHER-LATITUDE FAUNAS

Schott (1935) recognized higher-latitude faunas below the surface sediment in numerous short submarine cores from the tropical Atlantic. The lower fauna in the cores was differentiated from the modern fauna by the absence of *Globorotalia menardii* (d'Orbigny). Phleger *et al.* (1953) recognized several higher-latitude faunas separated by faunas normal for the latitude in 39 long cores collected from 0° Lat. to 42° N. Lat. in the Atlantic. "Cold water faunas" are recorded (Phleger, 1939, 1942) in numerous submarine cores collected from the continental slope off the northeastern United States, from 37° N. Lat. to 41° N. Lat. Additional reports from the Atlantic and other areas are: one core from the Caribbean Sea (Phleger, 1948); numerous cores from the northwestern Gulf of Mexico (Phleger, 1951b); numerous other cores from the Gulf of Mexico (Ewing *et al.*, 1958; Phleger, 1955b, 1960a); several cores collected by Ericson *et al.* (1952) from the western North Atlantic; in cores described by Bramlette and Bradley (1940) and Cushman and Henbest (1940) from approximately 50° N. Lat. in the Atlantic; in three cores from the equatorial Atlantic by Schott (1952). Stubbings (1937) studied six cores from the Arabian Sea and reports as many as four "colder water" faunas separated by "warm water" assemblages. A sequence of "cold" and "warm" faunas has been reported from the Tyrrhenian Sea (Phleger, 1947). Arrhenius (1952) recognized several Pleistocene stages in the eastern equatorial Pacific based in part on sequences of planktonic Foraminifera. Todd (1958) and F. L. Parker (1958) have reported Pleistocene assemblages from numerous long cores in the Mediterranean.

These "colder water" faunas have been correlated with glacial stages and/or substages of the Pleistocene, and the faunas normal for the latitude have been indicated as interglacial or inter-substage assemblages. Several "glacial stages" are reported by Schott (1952), Cushman and Henbest (1940), Phleger (1951b), Phleger and Hamilton (1946), Stubbings (1937), Todd (1958), F. L. Parker (1958), and others, and some of these authors have suggested that most or all of the

Pleistocene may be represented in some cores. Many of these interpretations are speculation and probably should be regarded as interesting possibilities, especially those involving low-latitude cores. We do not yet know enough about the effects of glaciation on the over-all distribution of marine water masses to interpret such sequences properly.

WATER MASS RELATIONSHIPS

There are at least two interesting reports of relationships of planktonic faunas to specific water masses and their probable geographic position during the Pleistocene. Arrhenius (1952) in his work on the eastern Pacific has shown that there is an abundance of tests of Foraminifera and other planktonic forms underlying the Equatorial Counter Current. In addition (*op. cit.*, p. 87) ". . . the results from the investigations of the east Pacific sequences . . . indicate great constancy in the path of the Counter Current during Pleistocene time. . . . The lack of significant displacements of the Counter Current during Pleistocene time appears to be proof that the climatic changes were synchronous in the northern and southern hemispheres."

It has been shown that the location of mixing of the Equatorial Counter Current and the Canaries Current can be recognized at 15-20° N. Lat. off west Africa by the mixed mid- and low-latitude faunas. In some of the long cores from the Atlantic there were lower assemblages, possibly Pleistocene in age, containing similar mixed faunas. These also occur near the coast of west Africa, but are 10-15° of latitude south of the present convergence, at 0-7° N. Lat. This strongly suggests a southward movement of the Equatorial Counter Current in the Atlantic during some time in the recent past, presumably Pleistocene. If this interpretation is correct, it has far-reaching implications. It appears that the position of the Northeast Trades must have been at least

10-15° south of the present position to cause such faunal mixing, and consequently the Counter Current must have been near the Equator instead of being located at 10-15° N. Lat. This suggests two further possibilities: either (1) the wind system and consequent oceanic current system were considerably "telescoped" during some part of the Pleistocene, or (2) the entire wind and current system was shifted to the southward into the southern hemisphere. If the latter possibility is correct, it may indicate that glaciation was not synchronous in the northern and southern hemispheres. Either the eastern Pacific and the eastern Atlantic behaved differently during the Pleistocene, or one of the above interpretations is incorrect. It is obvious that much more work is necessary before these interpretations can be evaluated.

PLEISTOCENE CORRELATIONS

Correlations of successions of the warm and cold water marine stages, based on Foraminifera in submarine cores, with stages of the European or the North American Pleistocene chronology have been attempted (see especially Emiliani, 1957). Such correlation is speculation, and should be considered only as an interesting possibility. One exception is the probable certainty of correlation of the fauna in the tops of cores above the last colder water fauna with at least much of the post-glacial or Holocene stage. Radiocarbon dates place the beginning of this at approximately 11,000-13,000 years ago (Emiliani, 1957; Ericson et al., 1956), a date which correlates very well with early post-glacial time in North America. It also seems reasonable that the colder-water fauna below the post-glacial is of Wisconsin-Würm age. Any other correlations with continental sequences are less reliable.

Emiliani (1957) apparently has correlated two Caribbean cores with the entire European Pleistocene sequence. Ericson and Wollin (1956), who have studied these same cores in considerable detail, on the other hand, have suggested that they may contain less than half of the Pleistocene sequence. Faunas in a long core (15 m.) from the Caribbean (Phleger, 1948) indicated numerous possible variations in surface-water temperatures. The writer speculated on the possibility of correlations of these faunas with the American and European Pleistocene sequences. Ovey (1949) reproduced an illustration attributed to Phleger (1948) in which the North American Pleistocene is correlated with the faunal sequences in the core. This illustration was copied in a textbook of paleontology by Sigal (1952, p. 294) and thereby has attained, by duplication, a certain credence.

Interpretations of stratigraphic sequences from ocean cores, whether biological, chemical, physical, or sedimentary, are in an early stage of development. Such interpretations should be accepted as working hypotheses but not as proved. Figures 81 and 82 illustrate interpretations (Phleger *et al.*, 1953) of foraminiferal faunas in twelve cores from the mid-latitudes of the North Atlantic and their suggested correlations. These are speculative and are to be considered with the qualifications mentioned. It does not seem desirable to attempt to assign age names to them based upon North American or European Pleistocene chronology.

Ericson and Wollin (1956) have given convincing evidence of correlation between cores in the equatorial Atlantic and the Caribbean. This is based on variations in sediment coarser than 74 microns and in direction of coiling of *Globorotalia truncatulinoides* (d'Orbigny) as well as on vertical distributions of planktonic faunas. Interpretations of planktonic foraminiferal faunas in thirteen cores from the northwest Gulf of Mexico are shown in Figure 83. These were taken along a traverse south of Trinity Shoal and southwest of Galveston, Texas. There probably is a plane of correlation between these cores, at the bottom of the modern fauna.

Recently the amount of post-glacial sedimentation in the

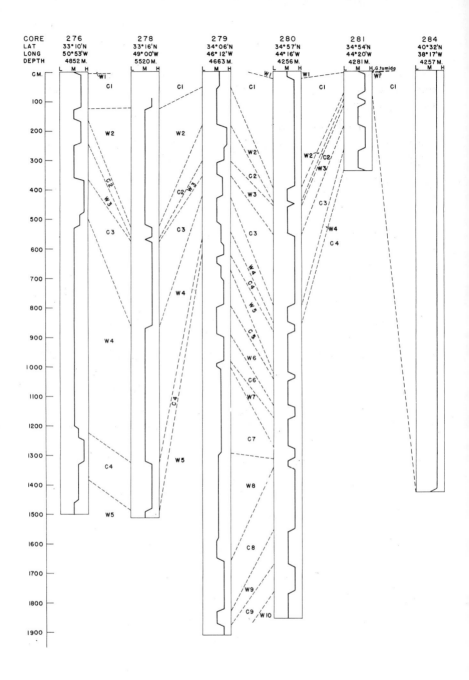

northern Gulf of Mexico has been tabulated by the thickness of the Recent planktonic fauna in 130 offshore cores (Phleger, 1960a). It is believed that the correlation of the boundary between these upper warm-water and the lower colder-water faunas is reliable in this area. The faunal change probably was contemporaneous throughout the basin since the introduction of North Equatorial Current water into the Strait of Yucatan would affect the entire Gulf at essentially the same time.

Most of the cores from the continental shelf did not penetrate the Recent planktonic fauna. Four cores did penetrate the post-glacial fauna, and these are all from near the edge of the shelf. The faunal boundary in these shelf cores ranges from 540 to 280 cm. In addition, the lower fauna in one core contains a benthonic fauna which suggests an inner shelf environment. In the northeastern area there is generally more than 150 cm. of Recent sediment on the upper continental slope and this decreases to less than 100 cm. in most cores deeper than 1000 fm. (1829 m.). On the central continental slope the Recent was not penetrated in any but one core from water deeper than 1000 fm. (1829 m.), and the Recent is thicker than approximately 200 cm. In the central Gulf of Mexico basin the sediment is about the same thickness as it is in the western basin. In the northeast there is generally less than 100 cm. on the upper slope, and in the basin the sediment is somewhat thicker.

There are two general implications of the studies of submarine cores. It is possible that (1) there was a southward

---

Figure 81, Opposite. Interpretations of faunas of planktonic Foraminifera in cores from the North Atlantic. L, low-latitude fauna; M, mid-latitude fauna; H, high-latitude fauna. C, colder-than-present; W, warm-as-present. After Phleger, Parker and Peirson (1953).

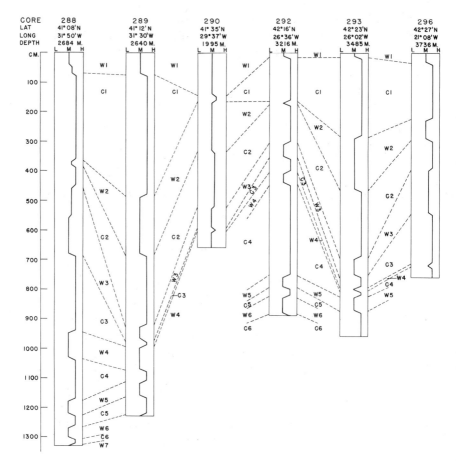

Figure 82. Interpretations of faunas of planktonic Foraminifera in cores from the North Atlantic. L, low-latitude fauna; M, mid-latitude fauna; H, high-latitude fauna. C, colder-than-present; W, warm-as-present. After Phleger, Parker and Peirson (1953).

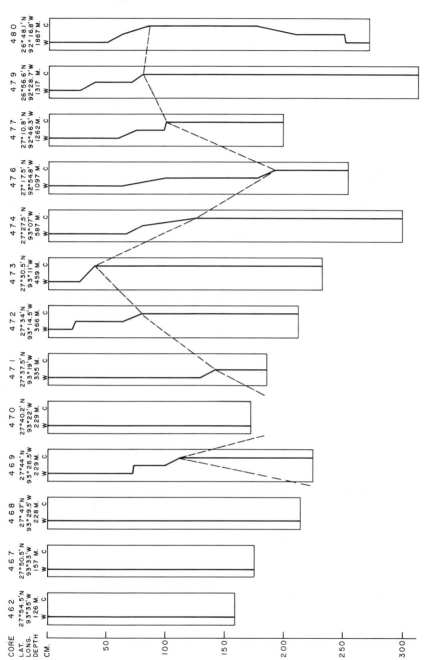

Figure 83. Interpretations of faunas of planktonic Foraminifera in cores from the northwestern Gulf of Mexico. W indicates relatively warm surface waters; C indicates relatively cold surface waters. After Phleger (1951b).

shift of the sea-surface isotherms during glacial stages, at least in the Atlantic, and (2) general cooling of the surface ocean water is indicated in the Atlantic during Pleistocene stages. There is insufficient knowledge of faunas in Pacific cores to make similar generalizations.

GEOLOGIC RANGES OF MODERN PLANKTONIC SPECIES

Many of the modern planktonic species range from the present back at least to the Miocene. One of the difficulties in assessing these geological ranges is the variety of names which have been used for some of the species. The considerable variation in most planktonic forms has tended to complicate their nomenclature. Many students of this group apparently have not considered variations within large suites of specimens, but have named certain key variants.

There is some evidence that *Globorotalia tumida* (Brady) has a limited range, at least in the Atlantic, probably extending back no further than the Middle Pleistocene and perhaps only to the last glacial stage. Many or most of the pre-modern specimens of *Globorotalia menardii* (d'Orbigny) are unlike the modern form and seem to be intermediate between *G. menardii* and *G. tumida*. This intermediate form appears not to exist at present and may have died out in the Pleistocene. It is our experience, however, that true *G. menardii* probably extends to the Miocene. All the modern species except *G. tumida* occur in the Miocene faunas recognized in deep-sea cores from the Atlantic. Some reports would extend a few of these species as far as the Eocene and would restrict others to the Pliocene (Ovey, *in* Wiseman and Ovey, 1950, Table 1). Ericson has suggested that *Globorotalia menardii flexuosa* (Koch) may be no younger than the last inter-glacial stage (Ericson *et al.*, 1956, p. 118).

Ericson *et al.* (1954) suggest that the direction of coiling of *Globorotalia truncatulinoides* (d'Orbigny) may have stratigraphic significance. Their observations in six cores show that in the majority of samples most of the specimens of this species have right coiling. Two layers in each core, however, have an unusual number of specimens of this species with left coiling, and there are indications that these can be correlated between the two cores. Systematic study of other variations in planktonic Foraminifera will probably provide further means for the correlation of post-Miocene faunas.

CHAPTER VI

Summary Discussion

SOME GENERALIZATIONS ON DISTRIBUTION

The data discussed in the preceding chapters are of only limited value in attacking problems of paleoceanography unless they can be organized into some logical system which may have general application. The matching of data on distribution of species probably may be used with success in the modern ocean and in the younger marine rocks of the geologic column. Any assemblage of animals is in reality composed of a series of individual, overlapping distributions, and it seems possible that no two distributions of species are identical. The reasons for such distributions are complex and may be somewhat different for different species. It is necessary, therefore, to have as much understanding as possible of distributions of different species and groups of species. It is also necessary to understand the variations within environments, insofar as this is possible.

The application of ecologic work on modern faunas to faunas in ancient marine sediments raises additional problems. Many or most older assemblages are not similar to the modern ones, as for example the Paleozoic faunas. It is important to attempt to discover how many of the present results may apply to such older faunal patterns. Another problem is the possibility that some forms occurring in older sediments may have had a different ecology from the same or similar forms living today.

A better understanding of the problems will be gained and ecologic results will be more useful if generalizations of distribution can be evolved. Such generalizations also form a series of working hypotheses which may be tested and serve as a guide to collection of additional data as well as additional theoretical considerations. The general statements listed below are based on the available information, and are subject to modification as new material is collected or present information is analyzed further.

1. Benthonic foraminiferal faunas are zoned offshore according to depth of water. There is a marked faunal boundary at the bottom of the seasonal layer. The depth of this in mid-latitudes varies from approximately 70-125 m. (38-68 fm.), depending upon the oceanography of the region in which it occurs. Other recognizable faunal depth boundaries on the continental shelf are at about 20-30 m. (11-16 fm.) and about 50 m. (27 fm.), and on the continental slope at 1000 m. and possibly at 2000 m. (547 and 1094 fm.). The boundaries between these depth zones are transitional and depths vary somewhat, but they seem to be world-wide in distribution based on data at present available.

2. Coastal lagoons have benthonic foraminiferal faunas which can be distinguished from the adjacent nearshore, open-ocean ones. Inner and lower lagoon assemblages can be differentiated where a distinctive inner lagoon water mass develops due to the size of the lagoon, high runoff or excessive evaporation.

3. Extensive marshes occur in areas of rapid alluviation, such as in a large delta or where a coastal lagoon is being

rapidly filled on the margins. Marsh faunas contaminate nearshore or lagoon faunas, and the extent of the contamination is a measure of the size of the marsh. Marsh faunas are similar in both brackish and hypersaline marshes.

4. Sand lagoon barriers may have a mixture of nearshore open-ocean, lagoon, and marsh benthonic Foraminifera. The open-ocean specimens may show a high degree of physical sorting where they are blown across such barriers along with sand.

5. Living-total ratios of benthonic Foraminifera may suggest relative rates of deposition of sediment and the size of an accumulated population may be an index to these. Very large populations of empty tests per unit amount of sediment suggest very slow detrital sedimentation; small populations suggest rapid deposition.

6. Experimental work (Bradshaw, personal communication) has shown that specimens of Foraminifera grown under optimum conditions where rapid reproduction occurs are generally much smaller in size than those cultured under less favorable conditions. In areas of high production such as off the Mississippi Delta (Lankford, 1959) where the conditions are optimum and support a very large living population, specimen size does *not* indicate a "depauperate" fauna, but indicates unusually favorable conditions and therefore rapid reproduction. In such areas of large living populations and small specimen size, rapid sedimentation rates often result in accumulating only a small population in the sediment. Specimens grown under marginal conditions reproduce very slowly and grow to larger-than-average size. Unusually large specimen size in Foraminifera may, therefore, indicate growth under marginal conditions.

7. "Natural contamination" between ecologically different benthonic assemblages is caused by:
    a. displacement of sediments into deeper water
    b. changes in depth of water due to rise and/or fall of sea-level

c. movement of currents and water masses

8. There are low-latitude, mid-latitude, and high-latitude assemblages of planktonic Foraminifera. This raises problems of long-range correlation by means of planktonic forms.

9. Mixed planktonic faunas in sediments occur in areas of marked convergence of water masses, where an element of the living fauna is relict from a previous environment, or where empty tests are residual from a previous environment.

10. Planktonic Foraminifera in abundance are characteristic of offshore, oceanic water masses. Populations generally may be low on the inner part of the continental shelf except where essentially oceanic water extends to the shore.

GENERAL ASPECTS OF FAUNAS

A more specific and perhaps more useful method of summarizing the information on ecology of Foraminifera is to consider certain general aspects of the faunas. Several population characteristics which may be considered are:
1. Ratio of planktonic to benthonic specimens
2. Number of benthonic species and genera
3. Per cent of the population composed of arenaceous specimens
4. Characteristic benthonic genera
5. Other special features of the fauna

The statements made below concerning these population characteristics should be considered approximations, and they apply only to areas of detrital sediments. They are based mostly on analyses of data from the Gulf of Mexico summarized in some of the earlier chapters and in part on other experience of the writer. Some of these data show considerable variation but it is believed that the generalizations given are good approximations of average values.

It should be stressed that no one of these characteristics used alone is necessarily a certain guide to recognizing a marine environment, but use of several should increase the reliability. Foraminifera are subject to complex biological and physical factors not yet well understood.

Marine marsh. All species are benthonic and average 5-8 species and 3-8 genera. Populations range from very low to very high. Arenaceous types comprise all or most of the population in most marshes; barrier sand island marshes may have more calcareous specimens than other types of marshes. The faunas seem to have world-wide similarity. Characteristic genera are:

>*Miliammina*
>*Arenoparrella*
>*Trochammina*
>*Ammoastuta*
>*Discorinopsis*
>*Jadammina*
>*Palmerinella*

Coastal lagoon. Faunas of lagoons are variable, depending upon runoff and tidal range, and the degree of difference from the nearby open-ocean fauna depends upon these factors. There are no planktonic forms or only occasional specimens. Population size is variable depending upon rate of sedimentation, and many lagoons average 5-10 genera and 10-20 species. Arenaceous specimens may comprise 5-75% of the population with more on inner brackish areas where these occur. The most characteristic genera in many lagoons may be *Ammotium*, *Streblus* and *Elphidium*. Numerous other genera are introduced from the open ocean; these are more abundant near inlets. There are fewer species in inner than in lower lagoons.

Nearshore turbulent zone (0-20 m. ±). The size of the population is variable but probably smaller per unit volume of sediment than in most or all other marine environments. Locally there are large populations. Planktonic forms usually are not present or there are only occasional specimens. The fauna is composed mostly of thick-shelled, large species of

Elphidium, Streblus, and Miliolidae; there may be large and robust arenaceous forms such as *Textularia.* The tests may be water-worn.

Inner continental shelf (20-60 m. ±). There are 10-25 benthonic species and 5-15 genera, generally fewer than on the outer shelf. 10-25% of the population may be arenaceous specimens. The planktonic-benthonic ratio may be less than 0.1 off coasts of moderately high runoff. The genera often having highest frequencies are *Buliminella, Rosalina, Elphidium, Rotalia,* and *Nouria.*

Outer continental shelf (60-100 m. ±). There are 30-40 species and 20-30 genera. The planktonic-benthonic ratio may be 0.1-1.0 with more planktonic species than on the inner shelf off coasts where there is appreciable runoff. Arenaceous specimens usually are approximately 5% of the benthonic population. The following genera may occur in relatively high frequencies: *Bigenerina, Cassidulina, Nonionella, Uvigerina, Virgulina, Cibicides,* and *Nonion.*

Upper continental slope (100-1000 m. ±). The planktonic-benthonic ratio is 1-5+. There are 30-40 species and 20-30 genera, and approximately 5% of the population is composed of arenaceous specimens. The following genera may be the most abundant: *Bolivina, Bulimina, Cassidulina, Pullenia,* and *Uvigerina.*

Lower continental slope and deep-sea (deeper than 1000 m.). The planktonic-benthonic ratio is approximately 10. There are approximately 10-25 species, 5-15 genera, and 10-20% arenaceous specimens in the benthonic population. Common genera are *Bulimina, Glomospira, Gyroidina, Haplophragmoides, Epistominella,* and *Pullenia.*

Comments on Geologic Applications

Possible application of the results on ecology of Foraminifera and the generalizations on distribution to older marine rocks deserves careful consideration. Application to middle and late Tertiary marine sediments is not expected to be complicated by significant differences in specific or generic composition. In early Tertiary sediments species assemblages closely related to those occurring in these environments have been found, and these facilitate facies definition. Comparison of Cretaceous faunas with modern ones is difficult and may be possible only in a general way.

Most of the modern species of benthonic Foraminifera extend back into geologic times as far as the Miocene, and direct comparisons may be possible between faunas of post-Miocene ages and the modern assemblages. Most modern species of planktonic Foraminifera also appear to extend to the Miocene but these are associated with fossil planktonic faunas and one or more modern species may extend into the Oligocene or even earlier. Many modern genera extend to Eocene and some to Lower Cretaceous. In the pre-Miocene and post-Jurassic faunas, therefore, interpretations of environments may possibly be based on associations of genera and/or species apparently closely related to modern forms.

An example of the use of families of Foraminifera for paleoecologic interpretation of Cretaceous marine sediments (the Grayson Marl of Texas) is given by Albritton *et al.* (1954, p. 327):

"Twenty-two samples collected from as many horizons of the Grayson were analyzed according to their content of foraminiferal tests, classified by genera and families. The Lituolidae are the predominant benthonic family in the upper clay. In the lower clay the Buliminidae generally outnumber the combined Anomalinidae, Rotaliidae, and Lagenidae, while in both marl units the converse is the case.

"Lowman (1949) has shown that different families of Foraminifera predominate at different depths beneath the present Gulf of Mexico. Comparing the microfaunal facies of the

Grayson with those described by him, it may be inferred that the lower two units of the Grayson were deposited in deepening waters, while the upper two were deposited in shoaling waters. This is compatible with the chronicle for the late Comanchean as known from studies of lithology, megafossils, and geologic structure."

In the upper clay of Albritton *et al.* (1954) there is an association of abundant Lituolidae and abundant Globigerinidae. In the present ocean the Lituolidae in relative abundance usually are characteristic of nearshore, low-salinity water; the Globigerinidae, on the other hand, are characteristic of offshore, oceanic-type water. This suggests the possibility that either the ecologic adjustment of one of the families has changed since the Cretaceous or the association reported is an artificial one. The possibility of change in ecologic adjustments of different groups of Foraminifera since the Jurassic or Cretaceous should not be overlooked in evaluation of older assemblages.

There are a few other applications of the results of ecology of modern Foraminifera to the ancient rocks, but a discussion of these is beyond the scope of this book. The use of families for this purpose, as Albritton *et al.* (1954) have done, and their use in defining modern faunas, as W. Moore (1955) has done, is considered to be a questionable method. The difficulty in the use of families for studies on ecology of Foraminifera lies in the differences of opinion about the contents of many families and different usages of this element of classification.

There is insufficient information and understanding now available to permit the preparation of a specific guide for the application of most ecological studies of modern Foraminifera for adequate paleoecologic interpretations of fossil sequences. It is desirable to make rather detailed studies of faunal patterns in attempting to apply the data on modern distribution to ancient marine sediments. The most satisfactory approach to the paleoecology of these organisms probably is by a progressive study from the best-known ecologies, *i.e.*, the present, to the least-known distributions, during the remote past. Much of our understanding of ecology will come from ecological studies of older faunas.

An older fauna should be analyzed in detail to obtain maximum information, both from an ecologic and a stratigraphic point of view. Faunal analysis probably should be quantitative where the material permits. Quantitative study may be impossible where there has been some destruction of the foraminiferal tests after deposition. Little is known about test destruction, but it may be selective so that some species are destroyed and others are unaffected. Fragments may be present which can be identified with sufficient knowledge of wall structure. It should be possible to estimate the fauna originally present from either fragments or a few whole tests. This requires detailed knowledge of the faunal associations and the presence of certain species. The technique of reconstructing a fauna from a few fragments has been demonstrated in Pleistocene planktonic faunas (Phleger *et al.*, 1953), illustrated on Plate 11.

It is desirable to attempt to reconstruct the ancient oceanography of a marine area as a background for predicting sedimentary patterns. This may be possible for some ancient nearshore areas where information is abundant. It appears desirable at the present stage of our knowledge to consider Foraminifera as "tracers" of different types of water, and as

---

Plate 11, Opposite. Upper illustrations show solution effects in *Globorotalia menardii* (d'Orbigny) and also the type of specimens present in deep-sea samples where solution of calcium carbonate has been marked. These are remnants of a fauna illustrated in the lower illustrations.

PRESENT IN SAMPLE

Pulleniatina obliquiloculata (Parker & Jones)

Globorotalia menardii (d'Orbigny)

FORMERLY PRESENT

Globigerina bulloides
d'Orbigny

Pulleniatina obliquiloculata
(Parker & Jones)

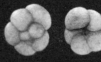

Globigerina eggeri
Rhumbler

Globigerinoides conglobata
(H.B. Brady)

Globigerinita glutinata
(Egger)

Globigerinoides sacculifera
(H.B. Brady)

Globorotalia punctulata
(d'Orbigny)

Globigerinoides rubra
(d'Orbigny)

Globigerinella aequilateralis
(H.B. Brady)

Globorotalia truncatulinoides
(d'Orbigny)

Globorotalia menardi
(d'Orbigny)

Globorotalia tumida
(H.B. Brady)

reflecting average oceanographic conditions. If the watermass distribution can be inferred it is then possible to infer the geography and the distribution of sedimentary processes, since these are closely interrelated. It is obvious that Foraminifera alone cannot give the entire analysis. It is necessary to understand characteristics of sedimentary materials other than Foraminifera, probable source of supply, *et cetera*. Other organisms having recognizable remains should have distribution patterns comparable to those of the Foraminifera.

In analyzing and reproducing an environmental complex such as a nearshore lagoon area with a barrier island, numerous samples with good distribution are desirable. Five or ten samples located at random and covering the entire area, however, may generally define the environments. Such information may suggest locations for additional data, or may be sufficient to make an intelligent estimate of the entire sedimentary environment and its distribution. It is desirable to have samples from actual exposures or well cores so that artificial faunal contamination can be avoided, but this is not always possible. Numerous important details which aid in reconstructing the patterns are based on faunal associations which are, in a sense, brought together by "natural contamination."

Guide fossils are a problem in beds having complex ecologic patterns. An effort should be made to select for guides species which occur in several environments. Differences in relative abundance may at times have ecologic, rather than stratigraphic significance. A difficulty may arise because of absence or rarity of a guide fossil in one or more environments, and the stratigraphic value thus may be limited. It may be more desirable to have a series of guide fossils each of which is abundant in one or more facies; their contemporaneity can be established by overlap ranges near facies boundaries where mixing occurs.

Frequencies of species and sizes of populations have been used by some workers in correlation of marine rocks, based on the assumption that similarities in these population parameters indicate contemporaneity. The general validity of this method for Foraminifera is questioned. It is believed that such a method of correlation could only be used in a single environment, and it is apparent that it cannot be used generally to correlate different environments. Where species occur in equal abundance in several environments the method could be used, but such distributions must be demonstrated. The danger in using frequencies for such a purpose is that correlation is sometimes made between similar ecologic suites of different ages and correlatives of different ecologies are concluded to be of different ages.

It was shown in Chapter IV that the total population of Foraminifera may be an index to the relative rate of deposition of sediment. Evidence is presented to suggest that a high population of benthonic Foraminifera per unit volume of sediment may indicate relatively slow deposition and a low population may indicate fast deposition. A high concentration of foraminiferal tests may be indicative of areas of "non-deposition," of "stratigraphic condensation," or of certain types of "unconformities." D. G. Moore (1954) has used the relative abundance of Foraminifera as a measure of rate of sedimentation in a core collected from San Antonio Bay, Texas. The higher concentrations of Foraminifera are interpreted by Moore as indicating slower rates of sedimentation. His work is based on the assumption that Foraminifera were produced at a uniform rate at the location of his core; this assumption is not proved but may be a reasonable approximation.

Oceanographic Applications

Some oceanographic applications of the studies on ecology of Foraminifera have been discussed in the preceding chapters. These are summarized and other possible applications are discussed briefly. It should be realized that several of these are somewhat speculative, but the present data indicate that they are possibilities which should be considered. The separation of oceanographic applications from geologic applications of these data is, in a sense, an artificial classification. Most or all of the discussion in this section also applies directly or indirectly to the geology of the ancient marine rocks, or paleoceanography.

Foraminifera occurring in sediments appear to make good "tracers" of the movement of sediment or of water masses where their normal habitat is known and is restricted, and they have been moved from this habitat into a foreign one. This may be regarded as a kind of "natural faunal contamination." The best-known example of this faunal contamination occurs in displaced sediments. A part or all of the microfauna moves with the coarse fraction of the sediment in this process and records movement of the sediment by being displaced by gravity from the normal depth of occurrence into deeper water. Some studies have suggested that faunal depth associations in such displaced sediment may be a clue to the method of transport downslope. The frequency and distribution of faunas displaced from their normal depth ranges will indicate the relative importance of this process in sedimentation in the deeper parts of the ocean.

It has also been shown that both benthonic and planktonic Foraminifera probably can be used to trace bodies of marine water which have a uniform ecologic effect. In nearshore areas such as lagoons where open-ocean water invades along the bottom but varies in position, occasional water observations do not necessarily give the best picture. Under these circumstances immigrant benthonic populations appear to invade with the bottom water and leave a record in the sediment of the presence of the water mass. The populations of benthonic Foraminifera in the sediment may be the best guide to the average position of such bottom water. This may be an aid to understanding the mechanisms of water exchange and water movement in lagoons and estuaries and provide a better generalization than actual observation and study of the water itself.

Different planktonic foraminiferal assemblages characterize different offshore oceanic water masses. Prevailing currents may introduce these forms into an area which is foreign to their normal habitat. The best-known example of this is the Gulf Stream which introduces low-latitude forms into mid-latitudes. These assemblages in sediments may be used to trace the average position of such a current and may indicate directions and amount of short-term and long-term migrations in position (see Boltovskoy, 1959a). It has been shown that convergence areas are marked by mixed planktonic faunas, and that there may have been a shift of the convergence area off Africa during the recent past. It appears possible that both short-term and long-term migrations of convergence areas might be analyzed by careful study of these mixed faunas. These data, along with the information about distribution of oceanic currents, should be useful in analyzing shifts in positions of climatic belts, for example, during the Pleistocene.

One type of natural contamination which may be of considerable importance in high latitudes is by ice rafting of sediments containing benthonic Foraminifera. In samples collected from Canadian and Greenland Arctic areas (Phleger, 1952a) considerable faunal mixing is obvious in the samples and it is suggested that this is caused by ice rafting. Faunal mixing and thus absence of clear-cut depth zonation may be a characteristic of Arctic areas, especially those areas in which there is considerable shallow water around an intricate coastline. Ice transportation of sedimentary materials, including Foraminifera, from nearshore zones and later deposition in deeper waters is probably one of the main sedimentary processes in Arctic and Antarctic regions.

The use of benthonic Foraminifera in determining rates of

marine sedimentation based on their rates of production is discussed in Chapter IV. It seems probable that planktonic Foraminifera also can be used for this, but they are more difficult to deal with because they form a more mobile population. There should be a definite proportion between the rate of production of Foraminifera and the total rate of organic production. It seems reasonable, therefore, that Foraminifera may eventually be useful in approximating organic production rates when these relationships are established.

There are many direct and indirect applications to processes and paleoceanography of the deep sea. Interpretation of climates and climatic changes and changes in circulation are of importance in deciphering deep-sea deposits and chronology. Possible depth changes, if of sufficient magnitude, may be recorded in benthonic forms. Areas of non-deposition or of erosion are recorded by high populations and fossil forms at the surface, as indicated for mid-Pacific seamounts by Hamilton (1953).

Foraminifera may aid in our understanding of the solution of calcium carbonate in the deep-sea. Abundant broken specimens were observed in Atlantic deep-sea faunas (Phleger *et al.*, 1953) and these were interpreted as being due principally to solution. This is based upon observations of whole specimens and fragments; a series of specimens believed to show progressive solution is shown in Plate 11. Some species appear to be more resistant to solution than others and may remain where others have been destroyed. The deep-sea benthonic species appear to be especially resistant to solution. The most intensive solution effects were seen in surface sediment samples from water depths greater than about 5000 m., but there are exceptions: a large population is recorded from 7500 m. Some cores have layers of broken specimens alternating with layers showing no evidence of solution.

Observations on solution of the calcareous faunas in cores

are of possible importance in studying relative rates of organic production, of deposition of sediment, and of chronology. Solution observations made on the Atlantic cores are not sufficient for definite applications. If most or all of the solution of calcium carbonate occurs at or near the sediment-water interface, then evidence at lower levels in a core should provide a clue to the conditions at the time when any fauna was deposited. Conditions causing intensive solution may have been a relatively low rate of Foraminifera production and perhaps other factors. It is possible that layers in the sediment containing intensive solution effects and also those containing fresh specimens are either to be correlated from area to area or have related causes. Such interpretations may be of value as a check on rates of deposition estimated from other sources, and as an aid in interpretations of past environments.

Arrhenius (1950a, p. 288) believes that these layers of broken foraminiferal tests in deep-sea cores are due to crushing by mud-eating organisms. He states: "It seems certain that the conditions mentioned indicate epochs when mud-eating organisms extensively reworked the sediment, crushing especially the larger and more fragile shells. The increased influence of the mud-eaters on the sediment may be caused either by a decreased rate of sedimentation or by an increased number of mud-eating organisms at the deep sea bottom. . . . If the rate of sedimentation is low, a layer will be reworked by organisms several times before being covered thickly enough to prevent the penetration of digging animals. . . . No traces of dissolution have hitherto been observed on entire shells in the zones of crushing. . . ."

Revelle (1944) summarized the distribution of calcium carbonate in the Pacific and commented on differences in Atlantic and Pacific water as follows (*op. cit.*, p. 126): "The fact that, in spite of the calcium brought into the Pacific from the Atlantic and Indian oceans by the deep water, there is a greater degree of undersaturation in the Pacific than in the Atlantic, can only mean that the rate of precipitation of $CaCO_3$ in waters near the surface is slower in the Pacific. This must be owing primarily to the much smaller amount of calcium-bearing river water which is emptied into the Pacific." Pettersson (personal communication) states that on the Swedish Deep-Sea Expedition "Alkalinity determinations were made for the first time in three oceans using the same technique, and give evidence of the important difference between the Atlantic and Pacific oceans with respect to saturation of carbonates. Whereas in the Atlantic saturation is established in all levels and regions investigated, with an over-saturation in the warm surface water to 40%, in the Pacific from the oxygen-minimum layer downward the water is under-saturated, and the over-saturation in the warm surface waters is smaller than in the Atlantic."

Arrhenius (1952) has pointed out that factors affecting accumulation rate of calcium carbonate are 1) rate of production of material, such as tests of Foraminifera, coccoliths, etc., and 2) rate of solution before burial. The latter is influenced by temperature, alkalinity, rate of circulation of water, and rate of deposition of non-calcareous material.

Research Problems

Several general research problems in the ecology of modern Foraminifera are suggested by the results which have been obtained in this field. These suggestions are those which appear to hold promise of results which will be useful in problems of oceanography and paleoceanography. The list is not complete and it will undoubtedly be modified as more understanding is gained of the problems in this field.

Study of variation in species (and faunas) is a field which holds considerable promise of useful results. The work of

Ericson *et al.* (1954) on the probable stratigraphic value of variation in *Globorotalia truncatulinoides* (d'Orbigny) is one of several examples of this type of work. It is possible that variations in other planktonic and benthonic species also may have stratigraphic significance. Variation as an indicator of environment is an interesting and promising possibility which warrants additional investigation.

An understanding of the significance of variation probably can be obtained only by physiological experimentation. Arnold (1954b) has shown that *Allogromia laticollaris* Arnold is quite variable in laboratory cultures. At room temperatures he has observed 0.3-2.6% of populations to show variations. Cultures kept at 14°C., however, showed more variation. No attempt was made to correlate variability with culture environments. In an earlier paper Arnold (1953) stated that variation in the number of apertures in this species probably is hereditary and does not appear to be affected by the environment.

John S. Bradshaw (personal communication) reports that the most normal appearing individuals of *Streblus beccarii* (Linné) were those cultured at low temperatures (10-15°C.) and with increase of temperature the chambers appeared to be more irregular. It is suggested by Bradshaw that this may be brought about by a relative scarcity of food because of increased metabolism at higher temperatures.

Abnormalities in individual specimens in foraminiferal populations from the Salton Sea, California, were studied by Arnal (1955). He reported the highest frequency of abnormal specimens on the edges of this body of water near the mouths of small streams. He also found abundant abnormal specimens in Playa del Rey Lagoon, California, a "confined and stagnant" body of water. The author concludes that abnormalities become abundant where the environment differs markedly from marine conditions. Bradshaw's experimental results suggest that where limits of tolerance are approached the individuals may become deformed.

Physiological experimentation is of great promise in testing the effect of various environmental factors on the survival, reproduction, and variation of different species, insofar as these experiments can be done under laboratory conditions. It should also be possible to determine rate of production and thus rate of supply of tests to sediment from such experiments. Laboratory experiments on the causes of very high production and the possible effect of trace materials merit consideration. Causes for high production in hypersaline coastal lagoons are especially puzzling. Experimental and field experience strongly indicate that small specimen size and few species are related to optimum conditions of the environment. Unusually large size appears to be characteristic of marginal environmental conditions. This suggests that the doctrine of the "depauperate" fauna, *i.e.*, small specimen size, few species, and relatively few specimens, may be an incorrect interpretation. Further experimentation and observation are desirable to clarify this problem. Results of experiments should be tested with natural distributions to place them in the proper perspective.

Much can be learned about the adaptations of these faunas by studying their distributions in the older rocks, and this is probably essential to an understanding of the problems. In this connection, it is desirable to attempt to use genera or other groups of species for ecological interpretation of pre-Miocene and post-Jurassic rocks. Species classified under different genera but similar in aspect to modern species may be useful.

Relatively little is known about the distribution of planktonic Foraminifera from plankton tows and almost any data are useful. It is of especial importance to know the bathymetric distribution of these types and their adjustments to various water masses. Such information is of possible usefulness in reconstructing conditions in the entire water column from accumulations of planktonic Foraminifera in the sediment. Seasonal, long-range determinations of the standing crop will aid in determining rate of production and thus rate of addition of planktonic tests to the sediment. Attempts to determine actual rate of settling under simulated marine conditions are desirable to know how well a sediment accumulation represents the water mass above it. The behavior of the tests under turbulent conditions and the effect of water stratification on rate of settling will be of interest.

It should be possible to learn much about the probable methods of movement of displaced sediments by study of the benthonic Foraminifera in areas where displacement occurs. This will require detailed analysis of the sediment, the Foraminifera, and the general oceanography in such an area.

There is insufficient information and understanding of depth zonation. Results obtained by Bandy (1954) and by the writer, discussed in Chapter II, suggest that there is a more detailed depth zonation on the continental shelf than has been realized. This should be investigated further in a variety of areas in an attempt to generalize on such distributions.

It is suggested in Chapter IV that there may be little sedimentation at present on the outer shelf in the Gulf of Mexico and elsewhere, on the basis of live-total population ratios. This is of considerable importance in our understanding of nearshore marine sedimentation and should be further investigated. Additional background work is necessary, however, to determine the significance of live-total population ratios of benthonic Foraminifera in estimating rate of sedimentation. Long-range, frequent sampling of the standing crop, supplemented by laboratory cultures, will aid in this problem. In addition, it is desirable to determine actual rate of organic production and the relationship of foraminiferal production to total organic production.

Examination of the trace element composition of foraminiferal tests will be of considerable value, both as an indication of which elements may be concentrated in the tests and because of possible correlation with environments of deposition of sediment. Studies reported by Said (1951) indicate that there are differences in amounts of the rarer elements in the shells of the same benthonic species (*Amphistegina radiata* Terquem) in the Red Sea and on Bikini Atoll in the Pacific. Emiliani (1955) has reported some geographic variations in the amounts of various elements in the shells of planktonic Foraminifera. Numerous other analyses must be made before any conclusive results are obtained. Experimental studies of the physiological effect of certain trace elements may be of interest.

It is the opinion of many geologists that the substrate has a more important effect on the distribution of Foraminifera than can at present be demonstrated. This should be investigated further, probably by experimentation.

The significance of mixing of sediment by organisms is understood only in a very general way. Observations suggest that Foraminifera may be useful in studying this sedimentary process by analyzing the faunal mixing which has occurred.

Detailed population patterns of subspecies, species, and larger groups deserve attention because of the application to species variation, to evaluation of distribution trends, and to understanding normal patterns in various taxonomic groups and environments. In this connection the mechanism of dispersal of benthonic forms is of great interest; dispersal may be by free-floating, young stages, as in numerous other marine invertebrates.

The evaluation of fragments of Foraminifera is of some importance in the interpretation of fossil faunas. Insight into this matter can be gained by artificially creating fragments of some modern assemblages.

Knowledge of nearshore distributions of Foraminifera is inadequate. Understanding of the environments in which these faunas occur is especially deficient; more study of the oceanography of such areas is required for further understanding of the distribution of the faunas. Study of nearshore areas in a variety of places, off coasts with different amounts of runoff and having different types of coastlines, is desirable for further understanding of this complex problem.

Study of sediments, and especially the faunas, from deep-sea, submarine cores is in its infancy. More detailed examination of faunas of Foraminifera and correlation with all the physical and chemical properties in such cores is recommended. More long cores should be studied from the Southern Hemisphere and from high latitudes in all oceans, and also from continental shelf areas.

The ecology of modern Foraminifera is only one phase of marine sedimentology, and this should be kept in mind by those who may do research in this field or attempt to apply the results. Knowledge of foraminiferal ecology has, nevertheless, been of great practical value in defining various marine environments and in recognizing pre-modern environments of deposition. Marine sediments contain materials derived both from the land and the ocean and there are complex relationships between the sediments, their sources, and the processes which have affected them. Significant progress will be made when all these factors are integrated. We are now on the threshold of understanding some of these relationships since many of the problems in this field of study are at least in part defined. Studies of modern Foraminifera have been made by specialists in this field and many have ignored other sedimentary materials or have no knowledge of the marine environment. Many students of marine sediments have either disregarded the Foraminifera and remains of other organisms, or have shown little understanding of how they should be evaluated. Many physical and chemical oceanographers who have studied features of the marine environment have little or no interest in any aspect of the sediment on the bottom. A unification of all these approaches will give a better understanding of the occurrences of different kinds of marine sediment and should lead to knowledge of the fundamental reasons for such occurrences. When this stage is attained we shall be in a position to interpret more accurately ancient marine sediments in terms of paleoceanography.

References

Akers, W. H., 1952, General ecology of the foraminiferal genus *Eponidella* with a description of a Recent species: Jour. Paleontology, v. 26, p. 645-649.

─────, 1954, Ecologic aspects and stratigraphic significance of the foraminifer *Cyclammina cancellata*: Jour. Paleontology, v. 28, p. 132-152.

Albritton, C. C., Jr., Schell, W. W., Hill, C. S. & Puryear, J. R., 1954, Foraminiferal populations in the Grayson Marl: Bull. Geol. Soc. America, v. 65, p. 327-336.

Arnal, R. E., 1955, Some occurrences of abnormal Foraminifera: The Compass, Sigma Gamma Epsilon, v. 32, p. 185-194.

Arnold, Z. M., 1953, Paleontology and the study of variation in living Foraminifera: Contr. Cushman Found. Foram. Research, v. 4, p. 24-26.

─────, 1954a, *Discorinopsis aguayoi* (Bermúdez) and *Discorinopsis vadescens* Cushman and Brönnimann; a study of variation in cultures of living Foraminifera: Contr. Cushman Found. Foram. Research, v. 5, p. 4-13.

─────, 1954b, Variation and isomorphism in *Allogromia laticollaris*: a clue to foraminiferal evolution: Contr. Cushman Found. Foram. Research, v. 5, p. 78-87.

Arrhenius, Gustaf, 1950a, Foraminifera and deep sea stratigraphy: Science, v. 111, no. 2881, p. 288.

─────, 1950b, Late Cenozoic climatic changes as recorded by the Equatorial Current System: Tellus, 2, p. 83-88.

─────, 1952, Sediment cores from the East Pacific: Repts. of the Swedish Deep-Sea Exped., 1947-1948, v. 5, fasc. 1, p. 6-91.

Bandy, O. L., 1953, Ecology and paleoecology of some California Foraminifera. Part 1. The frequency distribution of Recent Foraminifera off California: Jour. Paleontology, v. 27, p. 161-182.

―――――, 1954, Distribution of some shallow-water Foraminifera in the Gulf of Mexico: U. S. Geol. Surv., Prof. Paper 254-F, p. 125-141.

―――――, 1956, Ecology of Foraminifera in northeastern Gulf of Mexico: U. S. Geol. Surv., Prof. Paper 274-G, p. 179-204.

―――――, & Arnal, R. E., 1957, Distribution of Recent Foraminifera off the west coast of Central America: Bull. Am. Assoc. Petroleum Geol., v. 41, p. 2037-2053.

Bartenstein, Helmut & Brand, Erich, 1938, Die Foraminiferen-Fauna des Jade-Gebietes. 2, Foraminiferen der Meerischen und Brackischen Begirke des Jade-Gebietes: Senckenbergiana, bd. 20, p. 386-412.

Bé, A. W. H., 1959, Ecology of Recent planktonic Foraminifera: Micropaleontology, v. 5, p. 77-100.

Behm, H. J., & Grekulinski, E. F., 1958, The ecology of Foraminifera of Main and Richmond creeks, Staten Island, N. Y.: Proc. Staten Island Inst. Arts & Sci., v. XX, p. 53-66.

Blanc-Vernet, Laure, 1958, Les milieux sédimentaires litteraux de la provence occidentale (côte rocheuse). Relations entre la microfaune et la granulometrie de sédiment: Bull. de l'Inst. Oceanographie, no. 1112, 45 p.

Bolli, H. M., & Saunders, J. B., 1954, Discussion of some Thecamoebina described erroneously as Foraminifera: Contr. Cushman Found. Foram. Research, v. 5, p. 45-52.

Boltovskoy, Esteban, 1954a, Beobachtungen über Einfluss der Ernährung auf die Foraminiferenschalen: Pal. Zeitschr., v. 28, p. 204-207.

―――――, 1954b, Foraminíferos de la Bahia San Blas: Inst. Nac. de Invest. de las Cienc. Nat., Mus. Argentino de Cienc. Nat., Cienc. Geol., tomo 3, no. 4, p. 247-300.

―――――, 1954c, Foraminíferos del Golfo San Jorge: Inst. Nac. de Invest. de las Cienc. Nat., Mus. Argentino de Cienc. Nat., Cienc. Geol., tomo 3, p. 79-228.

―――――, 1955, Recent Foraminifera from shore sands at Quequén, Province of Buenos Aires, and changes in the foraminiferal fauna to the north and south: Contr. Cushman Found. Foram. Research, v. 6, p. 39-42.

―――――, 1956, Applications of chemical ecology in the study of Foraminifera: Micropaleontology, v. 2, p. 321-325.

―――――, 1957, Los Foraminíferos del Estuario del Rio de la Plata y su Zona de Influencia: Inst. Nac. de Invest. de las Cienc. Nat., Mus. Argentino de Cienc. Nat., Cienc. Geol., tomo 6, p. 1-77.

―――――, 1958, The foraminiferal fauna of the Rio de la Plata and its relation to the Caribbean fauna: Contr. Cushman Found. Foram. Research, v. 9, p. 18-21.

―――――, 1959a, Foraminifera as biological indicators in the study of ocean currents: Micropaleontology, v. 5, p. 473-481.

―――――, 1959b, Foraminiferos Recientes del sur de Brasil: Republica Argentina, Secretaria de Marina, Servicio de Hidrografia Naval, p. 1-124.

Bradshaw, J. S., 1955, Preliminary laboratory experiments on ecology of foraminiferal populations: Micropaleontology, v. 1, p. 351-358.

———, 1957, Laboratory studies on the rate of growth of the foraminifer, "*Streblus beccarii* (Linné) var. *tepida* Cushman": Jour. Paleontology, v. 31, p. 1138-1147.

———, 1959, Ecology of living planktonic Foraminifera in the north and equatorial Pacific Ocean: Contr. Cushman Found. Foram. Research, v. 10, p. 25-64.

Bramlette, M. N., & Bradley, W. H., 1940, Geology and biology of North Atlantic deep-sea cores between Newfoundland and Ireland. Pt. 1, Lithology and geologic interpretations: U. S. Geol. Surv., Prof. Paper 196-A, p. 1-34.

Butcher, W. S., 1951, Foraminifera, Coronado Bank and vicinity, California: Univ. of Calif., Los Angeles, Doctoral Thesis.

Carsola, A. J., 1952, Marine geology of the Arctic Ocean off Alaska and northwestern Canada: Univ. of Calif., Los Angeles, Doctoral Thesis.

Christiansen, Bengt, 1958, The foraminifer fauna in Drobak Sound in the Oslo Fjord (Norway): Nytt. Mag. f. Zool., v. 6, p. 5-91.

Clarke, G. L., & Bumpus, D. F., 1950, The plankton sampler—an instrument for quantitative plankton investigations: Am. Soc. Limnology & Oceanography, Special Pub. no. 5.

Colom, Guillermo, 1950, Estudio de los Foraminíferos de muestras de fondo recogidas entre los cabos Juby y Bojador: Bol. Inst. Español Oceanografia, no. 28, p. 1-45.

———, 1952, Foraminíferos de las costas de Galicia: Bol. Inst. Español Oceanografia, no. 51, p. 1-58.

Crouch, R. W., 1952, Significance of temperature on Foraminifera from deep basins off southern California coast: Bull. Am. Assoc. Petroleum Geol., v. 36, p. 807-843.

———, 1954, Paleontology and paleoecology of the San Pedro Shelf and vicinity: Jour. Sed. Petrology, v. 24, p. 182-190.

Curray, J. R., 1960, Sediments and history of the Holocene transgression, continental shelf, northwest Gulf of Mexico, *in* Recent sediments, northwestern Gulf of Mexico: Am. Assoc. Petroleum Geol., Special Pub. (in press).

Cushman, J. A., 1948, Foraminifera, their classification and economic use: 478 p., 55 pl., Cambridge, Harvard University Press.

———, & Henbest, L. G., 1940, Geology and biology of North Atlantic deep-sea cores between Newfoundland and Ireland. Pt. 2, Foraminifera: U. S. Geol. Surv., Prof. Paper 196-A, p. 35-50.

———, Todd, Ruth, & Post, R. J., 1954, Recent Foraminifera of the Marshall Islands: U. S. Geol. Surv., Prof. Paper 260-H, p. 319-384.

Drooger, C. W., & Kaasschieter, J. P. H., 1958, Foraminifera of the Orinoco-Trinidad-Paria shelf: Orinoco Shelf Exped., v. 4, Eerste Reeks, Deel 22, Amsterdam, p. 1-108.

Dryden, A. L., 1931, Accuracy in percentage representation of heavy mineral frequencies: Proc. Natl. Acad. Sci., v. 17, p. 233-238.

Emery, G. R., & Broussard, D. E., 1954, A modified Kullenberg piston corer: Jour. Sed. Petrology, v. 24, p. 207-211.

Emery, K. O., Butcher, W. S., Gould, H. R., & Shepard, F. P., 1952, Submarine geology off San Diego, California: Jour. Geology, v. 60, p. 511-548.

—————, Stevenson, R. E., & Hedgpeth, J. W., 1957, Estuaries and lagoons, *in* Treatise on marine ecology and paleoecology, v. 1, Ecology: Geol. Soc. America, Mem. 67, p. 673-751.

Emiliani, Cesare, 1954, Depth habitats of some species of pelagic Foraminifera as indicated by oxygen isotope ratios: Am. Jour. Sci., v. 252, p. 149-158.

—————, 1955, Mineralogical and chemical composition of the tests of certain pelagic Foraminifera: Micropaleontology, v. 1, p. 377-380.

—————, 1957, Temperature and age analysis of deep-sea cores: Science, v. 125, p. 383-387.

Ericson, D. B., Broecker, W. S., Kulp, J. L., & Wollin, Goesta, 1956, Late Pleistocene climates and deep-sea sediments: Science, v. 124, p. 385-389.

—————, Ewing, Maurice, & Heezan, B. C., 1951, Deep-sea sands and submarine canyons: Bull. Geol. Soc. America, v. 62, p. 961-965.

—————, 1952, Turbidity currents and sediments in North Atlantic: Bull. Am. Assoc. Petroleum Geol., v. 36, p. 489-511.

—————, & Wollin, Goesta, 1956, Correlation of six cores from the equatorial Atlantic and the Caribbean: Deep-Sea Research, v. 3, p. 104-125.

—————, Wollin, Goesta, & Wollin, Janet, 1954, Coiling direction of *Globorotalia truncatulinoides* in deep-sea cores: Deep-Sea Research, v. 2, p. 152-158.

Ewing, Maurice, Ericson, D. B., & Heezan, B. C., 1958, Sediments and topography of the Gulf of Mexico, *in* Habitat of oil: Am. Assoc. Petroleum Geol., p. 995-1053, Tulsa.

Flint, R. F., 1947, Glacial geology and the Pleistocene epoch: 589 p., New York, John Wiley & Sons.

Fuglister, F. C., & Worthington, L. V., 1951, Some results of a multiple ship survey of the Gulf Stream: Tellus, v. 3, p. 1-14.

Galloway, J. J., 1933, A manual of Foraminifera: 483 p., Bloomington, Ind.

Glaessner, M. F., 1945, Principles of micropaleontology: 296 p., New York, John Wiley & Sons.

Greenman, N. N., & Leblanc, R. J., 1956, Recent marine sediments and environments of northwest Gulf of Mexico: Bull. Am. Assoc. Petroleum Geol., v. 40, p. 813-847.

Grimsdale, T. F., & Van Morkhoven, F. P. C. M., 1955, The ratio between pelagic and benthonic Foraminifera as a means of estimating depth of deposition of sedimentary rocks: Proc. 4th World Petroleum Congr., Sect. I/D, Paper 4, p. 473-491.

Hamilton, E. L., 1953, Upper Cretaceous, Tertiary, and Recent planktonic Foraminifera from Mid-Pacific flat-topped seamounts: Jour. Paleontology, v. 27, p. 204-237.

Harvey, H. W., 1955, The chemistry and fertility of sea waters: 224 p., Cambridge, Cambridge Univ. Press.

Hedberg, H. D., 1934, Some Recent and fossil brackish to fresh-water Foraminifera: Jour. Paleontology, v. 8, p. 469-476.
Hedgpeth, J. W., 1957, Obtaining ecological data in the sea, in Treatise on marine ecology and paleoecology, v. 1, Ecology: Geol. Soc. America, Mem. 67, p. 53-86.
Höglund, Hans, 1947, Foraminifera in the Gullmar Fjord and the Skagerak: Zool. Bidrag Fran Uppsala, Bd. 26, p. 1-311.
Houbolt, J. J. H. C., 1957, Surface sediments of the Persian Gulf near the Qatar Peninsula: 113 p., The Hague, Mouton & Co.
Hutchins, L. W., 1947, The bases for temperature zonation in geographical distribution: Ecology Mon., 17, p. 325-335.
Hvorslev, M. J., & Stetson, H. C., 1946, Free-fall coring tube: a new type of gravity bottom sampler: Bull. Geol. Soc. America, v. 57, p. 935-950.
Illing, M. A., 1952, Distribution of certain Foraminifera within the littoral zone on the Bahama Banks: Ann. Mag. Nat. History, v. 5, p. 275-285.
Isaacs, J. D., & Maxwell, A. E., 1952, The ball-breaker, a deep water bottom signalling device: Jour. Marine Research., v. 11, p. 63-68.
Jarke, J., 1958, Sedimente und Microfaunen in Bereich der Grenzschwelle zweier oceanischer Räurne, dargestellt an einem Schmitt über den Island-Färöer-Ruchen: Sonder. Geol. Rund., Bd. 47, p. 234-249.
Kruit, Cornelis, 1955, Sediments of The Rhone delta: Mouton & Co., 's Gravenhage, p. 1-41.
Krumbein, W. C., & Garrels, R. M., 1952, Origin and classification of chemical sediments in terms of pH and oxidation-reduction potentials: Jour. Geology, v. 60, p. 1-33.
Kullenberg, Bjorge, 1947, The piston core sampler: Svenska Hydr. Biol. Kamm. Skr. 3, Ser. Hydr., v. 1, p. 1-46.
Lankford, R. R., 1959, Distribution and ecology of Foraminifera from east Mississippi Delta margin: Bull. Am. Assoc. Petroleum Geol., v. 43, p. 2065-2099.
Le Calvez, Jean, 1936, Modifications du test des Foraminifères pélagique en rapport avec le reproduction: *Orbulina universa* d'Orbigny et *Tretomphalus bulloides* d'Orbigny: Ann. Prot., v. 5, p. 125-133.
———, 1938, Recherches sur les Foraminifères: Arch. de Zool. Exper. et Gen., T. 80, p. 163-333.
———, & Le Calvez, Yolande, 1951, Contribution a l'étude des Foraminifères des eaux Soumâtres: Vie et Milieu, Tone II, fasc. 2, p. 237-254.
Lehmann, E. P., 1957, Statistical study of Texas Gulf Coast Recent foraminiferal facies: Micropaleontology, v. 3, p. 325-356.
Leroy, L. W., 1938, A preliminary study of the microfaunal facies along a traverse across Peper Bay, west coast of Java, De Ingenieur in Nederlandsch-Indie, IV. Mijnbouw en Geologie, de Mynengenieur, 5d.: Jaargang, no. 8, p. 130-133.
Loeblich, A. R., Jr., & Tappan, Helen, 1953, Studies of Arctic Foraminifera: Smithsonian Misc. Coll., v. 121, no. 7 (pub. 4105), p. 1-150.

Logan, B. W., 1959, Environments, foraminiferal facies and sediments of Shark Bay, Western Australia: Univ. of Western Australia, Nedland, Doctoral Thesis.

Lowman, S. W., 1949, Sedimentary facies in Gulf Coast: Bull. Am. Assoc. Petroleum Geol., v. 33, p. 1939-1997.

Ludwick, J. C., 1950, Deep water sand layers off San Diego: Univ. of Calif., Los Angeles, Doctoral Thesis.

McGlasson, R. H., 1959, Foraminiferal biofacies around Santa Catalina Island, California: Micropaleontology, v. 5, p. 217-240.

McKee, E. D., Chronic, J., & Leopold, E. B., 1959, Sedimentary belts in lagoon of Kapingamarangi Atoll: Bull. Am. Assoc. Petroleum Geol., v. 43, p. 501-562.

Moberg, E. G., 1927, Observations on tidal changes on physical and chemical conditions of sea water in the San Diego region: Bull. Scripps Inst. Oceanography, Tech. Ser., v. 1, p. 1-14.

Moore, D. G., 1954, Rates of deposition as indicated by Foraminifera abundance: Am. Petroleum Inst. Proj. 51, Rept. XIV, Inst. Marine Research, Univ. Calif., Ref. 54-1, p. 10-12.

Moore, W. E., 1957, Ecology of Recent Foraminifera in northern Florida Keys: Bull. Am. Assoc. Petroleum Geol., v. 41, p. 727-741.

Morishima, Masao, 1948, Foraminiferal thanatocoenoses of Ago Bay, Kii, Japan: Rept. of the Committee on a treatise on marine ecology and paleoecology, p. 111-117.

―――――, 1955, Deposits of foraminiferal tests in the Tokyo Bay, Japan: Mem. College of Sci., Univ. of Kyoto, Ser. B, v. 22, no. 2, p. 213-222.

―――――, & Chiji, Manzo, 1952, Foraminiferal thanatocoenosis of Akkeshi Bay and its vicinity: Mem. College of Sci., Univ. of Kyoto, Ser. B, v. 20, p. 113-117.

Munk, W. H., 1949, Surf beats: Trans. Am. Geophys. Union, v. 30, p. 849-854.

Murray, John, 1897, On the distribution of the pelagic Foraminifera at the surface and on the floor of the ocean: Nat. Science, v. 10, 65, p. 17-27.

Myers, E. H., 1935a, Culture methods for the marine Foraminifera of the littoral zone: Trans. Micr. Soc., v. LIV, no. 3, p. 264-267.

―――――, 1935b, The life history of *Patellina corrugata* Williamson, a foraminifer: Bull. Scripps Inst. Oceanography, Tech. Ser., v. 3, p. 355-392.

―――――, 1936, The life-cycle of *Spirillina vivipara* Ehrenberg, a foraminifer: Jour. Royal Micr. Soc., v. LVI, p. 120-146.

―――――, 1937, Culture methods for marine Foraminifera of the littoral zone, *in* Culture methods for invertebrate animals: p. 93-96, Ithaca, N. Y.

―――――, 1942, A quantitative study of the productivity of the Foraminifera in the sea: Proc. Am. Phil. Soc., v. 85, p. 325-342.

―――――, 1943, Biology, ecology, and morphogenesis of a pelagic foraminifer: Stanford Univ. Pub. in Biol. Sci., v. 9, no. 1, p. 5-30.

Natland, M. L., 1933, The temperature and depth distribution for some Recent and fossil Foraminifera in the southern California region: Bull. Scripps Inst. Oceanography, Tech. Ser., v. 3, no. 10, p. 225-230.

———, & Kuenen, Ph. H., 1951, Sedimentary history of the Ventura Basin, California, and the action of turbidity currents: Soc. Econ. Mineral. and Paleont., Special Pub. no. 2, p. 76-107.
O'Brien, M. P., 1931, Estuary tidal prisms related to entrance areas: Civil Engin., v. 1, no. 8, p. 738-739.
Oppenheimer, C. H., & ZoBell, C. E., 1952, The growth and viability of sixty-three species of marine bacteria as influenced by hydrostatic pressure: Sears Found., Jour. Marine Research, p. 10-17.
Otto, G. H., 1933, Comparative tests of several methods of sampling heavy mineral concentrates: Jour. Sed. Petrology, v. 3, p. 30-39.
Ovey, C. D., 1949, Note on the evidence for climatic changes from sub-oceanic cores: Weather, v. 4, p. 228-231.
Parker, F. L., 1948, Foraminifera of the continental shelf from the Gulf of Maine to Maryland: Bull. Mus. Comp. Zoology, v. 100, no. 2, p. 213-241.
———, 1952, Foraminiferal distribution in the Long Island Sound—Buzzards Bay area: Bull. Mus. Comp. Zoology, v. 106, no. 10, p. 427-473.
———, 1954, Distribution of the Foraminifera in the northeastern Gulf of Mexico: Bull. Mus. Comp. Zoology, v. 111, no. 10, p. 453-588.
———, 1955, Distribution of planktonic Foraminifera in some Mediterranean sediments, in Papers on Marine Biology and Oceanography: Deep-Sea Research, Supp. to v. 3, p. 204-211.
———, 1958, Eastern Mediterranean Foraminifera: Repts. Swedish Deep-Sea Exped., v. 8, fasc. 2, p. 217-283.
———, 1960, Living planktonic Foraminifera from the equatorial and southeast Pacific: (in press).
———, & Athearn, W. D., 1959, Ecology of marsh Foraminifera in Poponesset Bay, Massachusetts: Jour. Paleontology, v. 33, p. 333-343.
———, Phleger, F. B., & Peirson, J. F., 1953, Ecology of Foraminifera from San Antonio Bay and environs, southwest Texas: Cushman Found. Foram. Research, Special Pub. no. 2. 75 p.
Parker, R. H., 1956, Macro-invertebrate assemblages as indicators of sedimentary environments in east Mississippi Delta region: Bull. Am. Assoc. Petroleum Geol., v. 40, p. 295-376.
———, 1960, Ecology and distribution patterns of marine macro-invertebrates, northern Gulf of Mexico, in Recent sediments, northwestern Gulf of Mexico: Am. Assoc. Petroleum Geol., Special Pub. (in press).
Phleger, F. B, 1939, Foraminifera of submarine cores from the continental slope: Bull. Geol. Soc. America, v. 50, p. 1395-1422.
———, 1942, Foraminifera of submarine cores from the continental slope. Part 2.: Bull. Geol. Soc. America, v. 53, p. 1073-1098.
———, 1945, Vertical distribution of pelagic Foraminifera: Am. Jour. Science, v. 243, p. 377-383.
———, 1947, Foraminifera of three submarine cores from the Tyrrhenian Sea: Goteborgs Kungl. Vetensk. och Vitterhets-Samhalles Hand., Sjatte Foljden, Ser. B, no. 5, p. 1-19.

―――, 1948, Foraminifera of a submarine core from the Caribbean Sea: Goteborgs, Kungl. Vetensk. och Vitterhets-Samhalles Hand., Sjatte Foljden, Ser. B, bd. 5, no. 14, p. 1-9.

―――, 1949, The Boylston Street fishweir II, The Foraminifera: Papers of the R. S. Peabody Found. for Archeology, v. 4, no. 1, p. 99-108.

―――, 1951a, Displaced Foraminifera faunas: Soc. Econ. Paleont. and Mineral., Special Pub. no. 2, p. 66-75.

―――, 1951b, Ecology of Foraminifera, northwest Gulf of Mexico. Pt. 1, Foraminifera distribution: Geol. Soc. America, Mem. 46, p. 1-88.

―――, 1952a, Foraminifera distribution in some sediment samples from the Canadian and Greenland Arctic: Contr. Cushman Found. Foram. Research, v. 3, p. 80-89.

―――, 1952b, Foraminifera ecology off Portsmouth, New Hampshire: Bull. Mus. Comp. Zoology, v. 106, no. 8, p. 315-390.

―――, 1954a, Ecology of Foraminifera and associated micro-organisms from Mississippi Sound and environs: Bull. Am. Assoc. Petroleum Geol., v. 38, p. 584-647.

―――, 1954b, Foraminifera and deep-sea research: Deep-Sea Research, v. 2, p. 1-23.

―――, 1955a, Ecology of Foraminifera in southeastern Mississippi Delta area: Bull. Am. Assoc. Petroleum Geol., v. 39, p. 712-752.

―――, 1955b, Foraminiferal faunas in cores offshore from the Mississippi Delta, in Papers on Marine Biology and Oceanography: Deep-Sea Research, supp. to v. 3, p. 45-57.

―――, 1956, Significance of living foraminiferal populations along the central Texas coast: Contr. Cushman Found. Foram. Research, v. 7, p. 106-151.

―――, 1960a, Sedimentary patterns of Foraminifera, northern Gulf of Mexico, in Recent sediments, northwestern Gulf of Mexico: Am. Assoc. Petroleum Geol., Special Pub (in press).

―――, 1960b, Foraminiferal populations in Laguna Madre, Texas: (in press).

―――, & Ewing, G. C., 1960, Sedimentology and oceanography of some Mexican coastal lagoons: (in press).

―――, & Hamilton, W. A., 1946, Foraminifera of two submarine cores from the North Atlantic basin: Bull. Geol. Soc. America, v. 57, p. 951-966.

―――, & Lankford, 1957, Seasonal occurrences of living Foraminifera in some Texas bays: Contr. Cushman Found. Foram. Research, v. 8, p. 93-105.

―――, & Parker, F. L., 1951, Ecology of Foraminifera, northwest Gulf of Mexico. Part II, Foraminifera species: Geol. Soc. America, Mem. 46, p. 1-64.

―――, Parker, F. L., & Peirson, J. F., 1953, North Atlantic core Foraminifera: Repts. Swedish Deep-Sea Exped., v. VII, no. 1, p. 1-122.

―――, & Walton, W. R., 1950, Ecology of marsh and bay Foraminifera, Barnstable, Mass.: Am. Jour. Science, v. 248, p. 274-294.

Polski, William, 1959, Foraminiferal biofacies off the north Asiatic coast: Jour. Paleontology, v. 33, p. 569-587.

REFERENCES

Pratje, Otto, 1931, Die Sediments der Deutsche Bucht, Eine Regional-Statische Untersuchung: Wiss. Meeresunt., N. F. Abt. Helgoland, 18, H. 2, Abh. 6, Kiel n. Leipzig.

Reiter, Martin, 1959, Seasonal variations in intertidal Foraminifera in Santa Monica Bay, California: Jour. Paleontology, v. 33, p. 606-630.

Resig, J. M., 1958, Ecology of Foraminifera of the Santa Cruz Basin, California: Micropaleontology, v. 4, p. 287-308.

Revelle, R. R., 1944, Marine bottom samples collected in the Pacific Ocean by the CARNEGIE on its seventh cruise, in Scientific results of Cruise VII of the CARNEGIE: Carnegie Inst. of Washington, Pub. 556, p. 1-183.

Robins, T. M., 1933, Maintenance and improvement of entrance channels to the Pacific Coast ports: World Ports, v. 21, no. 3, p. 18-21.

Rottgardt, Dietrich, 1952, Mikropaläeontologisch Wichtige Bestandteile recenter brackische Sedimente an den Küsten Schleswig-Holsteins: Geol. Inst. Univ. Kiel, Bd. 1, p. 169-228.

Ruscelli, Maria, 1949, Foraminiferi de due saggi di fondo del Mar Ligure: Atti Accad. Ligure Sci. Lett., v. 6, fasc. 1, p. 1-31.

Rusnak, G. A., 1960, Sediments of Laguna Madre of Texas, in Recent sediments, northwestern Gulf of Mexico: Am. Assoc. Petroleum Geol., Special Pub. (in press).

Said, Rushdi, 1950, The distribution of Foraminifera in the northern Red Sea: Contr. Cushman Found. Foram. Research, v. 1, p. 9-29.

_____, 1951, Preliminary note on the spectroscopic distribution of elements in the shells of some Recent calcareous Foraminifera: Contr. Cushman Found. Foram. Research, v. 2, p. 11-13.

Saidova, H. M., 1957a, Quantitative distribution of Foraminifera in the Okhotsk Sea: Doklady Akad. Nauk SSSR, tom 114, no. 6, p. 1302-1305.

_____, 1957b, On the distribution of the Foraminifera in the strata of the sediments of the Okhotsk Sea: Doklady Akad. Nauk SSSR, tom 115, no. 6, p. 1213-1216.

_____, 1958, New data on the ecology of Foraminifera: Priroda, Oktjabr', Ottisk 12 No. 10, p. 107-110.

Schott, Wolfgang, 1935, Die Foraminiferen in den Aquatorialen Teil des Atlantischen Ozeans: Deutsche Atlantische Exped. 11, Heft 6, p. 411-616.

_____, 1952, On the sequence of deposits in the Equatorial Atlantic Ocean: Goteborgs Kungl. Vetensk. och Vitterhets-Samhalles Handl., Sjatte Foljden, Ser. B, Bd. 6, no. 2, p. 1-15.

Scruton, P. C., 1953a, Deposition of evaporites: Bull. Am. Assoc. Petroleum Geol., v. 37, p. 2498-2512.

_____, 1953b, Depth changes, Breton Sound to Pass a Loutre: Am. Petroleum Inst. Rept. no. VII, Scripps Inst. Oceanography Ref. 53-1, p. 15, fig. 7.

_____, 1956, Oceanography of Mississippi Delta sedimentary environments: Bull. Am. Assoc. Petroleum Geol., v. 40, p. 2864-2952.

Shenton, E. H., 1957, A study of the Foraminifera and sediments of Matagorda Bay, Texas: Trans. Gulf Coast Assoc. Geol. Soc., v. VII, p. 135-150.

Shepard, F. P., 1953, Sedimentation rates in Texas estuaries and lagoons: Bull. Am. Assoc. Petroleum Geol., v. 37, p. 1919-1934.

———, 1956, Marginal sediments of Mississippi Delta: Bull. Am. Assoc. Petroleum Geol, v. 40, p. 2537-2623.

———, & Lankford, R. R., 1959, Sedimentary facies from shallow borings in lower Mississippi Delta: Bull. Am. Assoc. Petroleum Geol., v. 52, p. 2051-2067.

Sigal, Jacques, 1952, Traité de paléontologie: Paris, p. 133-178; 192-301.

Stainforth, R. M., 1952, Ecology of arenaceous Foraminifera: The Micropaleontologist, v. VI, no. 1, p. 42-43.

Stetson, H. C., 1949, The sediments and stratigraphy of the East Coast continental margin; Georges Bank to Norfolk Canyon: Papers in Phys. Oceanography and Meteorology, Mass. Inst. Tech. and Woods Hole Oceanogr. Inst., v. XI, no. 2, p. 1-60.

———, 1953, The sediments of the western Gulf of Mexico. Part 1, The continental terrace of the western Gulf of Mexico: its surface sediments, origin and development: Papers in Phys. Oceanography and Meteorology, Mass. Inst. Tech. and Woods Hole Oceanogr. Inst., v. XIII, no. 4, p. 1-45.

———, & Parker, F. L., 1942, The Boylston Street fishweir, mechanical analysis of the sediments and identifications of the Foraminifera from the building excavation: Papers of the R. S. Peabody Found. for Archeology, v. 2, p. 41-44.

Stevenson, R. E., 1954, The marshlands at Newport Bay, California: Univ. S. Calif., Doctoral Thesis.

Stewart, H. B., Jr., 1958, Sedimentary reflections of depositional environment in San Miguel Lagoon, Baja California, Mexico: Bull. Am. Assoc. Petroleum Geol., v. 42, p. 2567-2618.

Stschedrina, Z. G., 1936, On the Foraminifera fauna of the Arctic seas of the USSR: Trudy Arkticheskogo Inst., tom 33, p. 51-64.

———, 1947, On the distribution of Foraminifera in the Greenland Sea: Doklady Akad. Nauk SSSR, tom 55, no. 9, p. 871-874.

———, 1957, Some regularities in the distribution of Recent Foraminifera: Trudy Leningradskogo Obshch. Estest., tom 73, no. 4, p. 99-106.

———, 1958a, The dependence of the distribution of Foraminifera in the seas of the USSR on the environmental factors: XV Int. Congr. Zool., Sect. III, Paper 30, p. 1-3.

———, 1958b, Foraminifera faunas of the eastern section of the Antarctic: Bull. Sovets. Antarktich. Eksped., no. 3, p. 51-54.

———, 1958c, Foraminifera Fauna of the seas around the southern Sakhalin and the southern Kuril Islands; explorations of the far eastern seas of the USSR, Issue V: Trudy Kurilo-Sakhalinskoj Krped., no. 1, p. 6-41.

———, 1958d, Foraminifera of the eastern Murman Sea: Trudy Murmansk. Biol. Sta., v. 4, p. 118-129.

———, 1958e, On the Foraminifera fauna of the Kuril-Kamtschatka trench: Trudy Inct. Okeanologii, Akad. Nauk SSSR, v. 27, p. 161-179.

Stubbings, H. G., 1937, Stratification of biological remains in marine deposits: The John Murray Exped., Sci. Repts., v. III, p. 154-192.

REFERENCES

Sverdrup, H. U., Johnson, M. W., & Fleming, R. H., 1942, The oceans: 1087 p., New York, Prentice-Hall.

Takayanagi, Yokichi, 1955, Recent Foraminifera from Matsukawa-Ura and its vicinity: Contr. Inst. Geology and Paleontology, Tohoku Univ., p. 1-96.

Thomas, W. H., & Simmons, E. G., 1960, Phytoplankton production in the Mississippi Delta, in Recent sediments, northwestern Gulf of Mexico: Am. Assoc. Petroleum Geol., Special Pub. (in press).

Thompson, E. F., 1939, The general hydrography of the Red Sea: The John Murray Exped., Sci. Repts., v. II, p. 83-103.

Todd, Ruth, 1958, Foraminifera from western Mediterranean deep sea cores: Repts. Swedish Deep-Sea Exped., v. VIII, fasc 2, p. 167-216.

―――――, & Brönnimann, Paul, 1957, Recent Foraminifera and Thecamoebina from the eastern Gulf of Paria: Cushman Found. Foram. Research, Special Pub. 3, 43 pp.

Trask, P. D., 1953, The sediments of the western Gulf of Mexico. Part II, Chemical studies of sediments of the western Gulf of Mexico: Papers in Phys. Oceanography and Meteorology, Mass. Inst. Tech. and Woods Hole Oceanogr. Inst., v. XII, no. 4, p. 49-120.

Treadwell, R. C., 1955, Sedimentology and ecology of southeast coastal Louisiana, in Trafficability and navigability of Louisiana coastal waters: Tech. Rept. no. 6, La. State Univ., Baton Rouge.

Uchio, Takayasu, 1960. Ecology of living benthonic Foraminifera from the San Diego, California, area: Cushman Found. Foram. Research, Special Pub. no. 5, p. 1-72.

Van Andel, Tj. H., Postma, H., et al., 1954, Recent sediments of the Gulf of Paria: Repts. of the Orinoco Shelf Exped., v. I, Verhandel. Kon. Nederl. Akad. Van Wetensch. Afd. Natuurkunde, Eerste Reeks, Deel XX, no. 5, 245 p.

Van Voorthuysen, J. H., 1951, Recent (and derived upper Cretaceous) Foraminifera of the Netherlands Wadden Sea (tidal flats): Meded. v. d. Geol. Sticht., N. Ser. no. 5, p. 23-32.

Walton, W. R., 1952, Techniques for recognition of living Foraminifera: Contr. Cushman Found. Foram. Research, v. III, p. 56-60.

―――――, 1955, Ecology of living benthonic Foraminifera, Todos Santos Bay, Baja California: Jour. Paleontology, v. 29, p. 952-1018.

Warren, A. D., 1956, Ecology of Foraminifera of the Buras-Scofield bayou region, southeast Louisiana: Trans. Gulf Coast Assoc. Geol. Soc., v. 6, p. 131-152.

Wiseman, J. D. H., & Ovey, C. D., 1950, Recent investigations on the deep-sea floor: Proc. Geol. Assoc., v. 61, p. 28-84.

ZoBell, C. E., 1946, Studies on redox potentials of marine sediments: Bull. Am. Assoc. Petroleum Geol., v. 30, p. 477-513.

―――――, & Johnson, F. H., 1949, Some effects of hydrostatic pressure on the multiplication and morphology of marine bacteria: Jour. Bacteriology, v. 60, p. 771-781.

Index

Adercotryma glomeratus, pl. 4, fig. 8; pl. 6, fig. 3
Ago Bay, Japan, faunal composition, 173-4
Akkeshi Bay, Japan, faunal composition, 174
Alveolophragmium
 crassimargo, pl. 6, fig. 4a, b
 jeffreysii, pl. 6, figs. 8, 9
 cf. *A. nitida,* pl. 6, figs. 10, 11
Ammoastuta inepta, pl. 8, figs. 11, 12
Ammobaculites
 cf. *A. dilatatus,* pl. 8, fig. 1
 sp. b, pl. 4, fig. 3
Ammoscalaria
 fluvialis, pl. 8, fig. 13
 pseudospiralis, pl. 1, fig. 2
 tenuimargo, pl. 4, fig. 2
Ammotium
 in Mississippi Sound, 134, 141
 salsum, pl. 9, figs. 4, 13
  living in San Antonio Bay area, fig. 64 (207)
 salsum variants, pl. 8, figs. 2, 7
Angulogerina
 angulosa, pl. 5, fig. 1
 bella, pl. 1, fig. 4
Arabian Sea, core studies, 244
Arctic, faunal boundaries, 42-3

Arenaceous species, destruction, 38
Arenoparrella mexicana, pl. 8, figs. 14-16
Asia coast
 depth distribution of benthonics, 40, 42-3
 nearshore distribution studies, 172-5
Asia continental shelf, planktonic species distribution, 241
Asterigerina carinata, pl. 1, fig. 3
Astrononion stellatum, pl. 5, fig. 2
Atlantic
 common planktonic species, pl. 10
 displaced faunas, 96-7
 distribution of *Globigerina bulloides,* fig. 71 (277)
 distribution of *Globorotalia menardii,* fig. 70 (226)
 generalized planktonic faunas, fig. 72 (228)
 interpretation of planktonic faunas in cores, figs. 81-2 (248, 250)
 mean monthly surface temperatures, fig. 1 (6)
 planktonic species distribution, 216-7
 Pleistocene correlation in cores, 247
 temperature variation, 4-5

Atlantic continental slope, depth distributions, 40-1

Baffin Bay, depth boundaries, 42
Bahama Banks, faunas, 184-5
Baja California coast
  depth distribution of benthonic species, 40,42
  evidence for lowered sea-level, 102
  sedimentation rate, 201
Barnstable Harbor, Mass.
  benthonic faunas, 126; fig. 47 (127)
  interpretation of faunas, 145
Beach and nearshore area
  general features, 15-16
  species, 158-9; pl. 9
Bengal, Bay of
  lowered sea-level in, 102
  sedimentation rate, 201
Bigenerina irregularis, pl. 1, fig. 1; pl. 7, figs. 1, 2
Bikini Atoll, fauna, 181-3
Biologic effects on distribution, 118, 120
Bolivina
  albatrossi, pl. 3, fig. 1
  barbata, pl. 2, fig. 1
  fragilis, pl. 2, fig. 2
  goesii, pl. 2, fig. 3
  lowmani, pl. 7, fig. 3; pl. 9, fig. 10
  minima, pl. 3, fig. 2
  ordinaria, pl. 3, fig. 4
  pseudoplicata, pl. 5, fig. 4; pl. 7, fig. 4
  pulchella primitiva, pl. 7, fig. 6
  subaenariensis mexicana, pl. 2, fig. 4
  translucens, pl. 3, fig. 3
  variabilis, pl. 7, fig. 5
Buccella
  frigida, pl. 5, fig. 14a, b
  hannai, pl. 1, fig. 7; pl. 7, fig. 9a, b
Bulimina
  aculeata, pl. 3, fig. 18; pl. 5, fig. 5
  alazanensis, pl. 3, fig. 20
  marginata, pl. 2, fig. 5
  spicata, pl. 3, fig. 19
Buliminella
  cf. *B. bassendorfensis,* pl. 1, fig. 8; pl. 7, fig. 7; pl. 9, fig. 8
  elegantissima, pl. 5, fig. 3; pl. 7, fig. 8

Calcium carbonate solution, 268-70
California coast, benthonic species depth distribution, 40,42
Cape Cod Bay, depth boundaries, 41
Caribbean Sea, core studies of planktonic species, 244
Cassidulina
  algida, pl. 5, fig. 6
  curvata, pl. 3, fig. 11
Centropyxis (Centropyxis) sp., pl. 8, figs. 20, 30
Chemical effects on distribution, 111-13
Chilostomella oolina, pl. 3, fig. 9
Cibicides
  deprimus, pl. 2, fig. 17
  io, pl. 1, fig. 9
  kullenbergi, pl. 4, figs. 17, 22
  lobatulus, pl. 5, fig. 8a, b; pl. 7, fig. 10a, b
  rugosa, pl. 3, fig. 17
  umbonatus, pl. 3, figs. 5, 6
  wuellerstorfi, pl. 4, figs. 21, 25
Cibicides lobatulus, distribution in Gulf of Maine, fig. 45, (119)
Clarke-Bumpus plankton sampler, 27, 30; fig. 10 (31)
Coastal lagoons
  description of environment, 13-15
  general aspects of fauna, 258
  hypersaline, 175
  Texas assemblages, 129, 133
  water masses, fig. 54 (140)
Coiling direction, *Globorotalia truncatulinoides,* 247
Continental shelf
  Asia, 172-5, 241
  Atlantic depth ranges, figs. 29, 30 (65,66)
  Atlantic distribution of benthonic species, 40-1
  depth ranges, northern Gulf of Mexico, 53; figs. 18-21 (54-57)
  general description, 17-18
  inner shelf fauna, 259
  inner shelf species, northern Gulf of Mexico, 80
  nature of seasonal layer, 17
  occurrence of planktonic species, northern Gulf of Mexico, fig. 80 (240)

INDEX 291

outer shelf fauna, 259
outer shelf species, northern Gulf of Mexico, 80
planktonic populations, 239, 241
sedimentation rates off Mississippi Sound, 191-2
Continental slope
  benthonic species depth distributions in Atlantic, 40, 41
  description of environment, 18
  fauna of upper slope, 259
Convergence areas of planktonic faunas, 225-8
Core studies of planktonic species
  Arabian Sea, 244
  Caribbean Sea, 244
  Gulf of Mexico, 244
  Mediterranean Sea, 244
  Tyrrhenian Sea, 244
Coring tube, description and operation, 21-23; fig. 5 (22)
Crithonina pisum hispida, pl. 6, fig. 1
Cyclammina cancellata, depth restriction, 52

Deep sea environment, 18
Deltaic marine fauna, species composition, 160-1; pl. 9
"Depauperate fauna", 272
  causes for, 256
  continental shelf off Argentina, 112
Depth assemblages, summary of world-wide, fig. 13 (44)
Depth boundaries, interpretation, 122-4
Depth distribution of species
  generalized local, 52-3, 80-2, 88-90
  world-wide, 47, 52
Depth range of species
  Elphidium spp., 102
  general, 43, 45
  Gulf of Maine, fig. 31 (67); pls. 5, 6
  Maine to Maryland, figs. 29, 30 (65,66)
  Northern Gulf of Mexico, figs. 14-17 (48-51), 18-21 (54-57), 22-28 (58-64); pls. 1-4
  Streblus beccarii var. A, 102
  Todos Santos Bay living, figs. 32-36 (83-87)

Difflugia
  urceolata, pl. 8, fig. 19
  sp., pl. 8, fig. 23
"Discorbis" squamata, pl. 5, fig. 10a, b; pl. 7, fig. 16a, b
Displaced faunas
  depth of origin, 95
  general, 90-99
  in Atlantic, 96-97
  in cores from San Diego Trough, 90-2; figs. 37-9 (91,93)
  in Pacific, 98
  in Sigsbee Deep cores, 92-5, fig. 40 (94)
  movement of, 92
  related to submarine canyons, 99

East China Sea, 43
Ecologic water masses
  effect on planktonic species, 242-3
  examples, 10
  importance of salinity in distinguishing, 109
  in marginal marine areas, 138
Eggerella advena, pl. 5, fig. 7
Ehrenbergina spinea, pl. 2, fig. 8
Elphidium
  articulatum, pl. 5, fig. 12
  delicatulum, pl. 8, fig. 5; pl. 9, fig. 11
  discoidale, pl. 7, fig. 17; pl. 9, fig. 18
  gunteri, pl. 7, fig. 18; pl. 9, figs. 1, 17
  incertum clavatum, pl. 5, fig. 11
  incertum mexicanum, pl. 7, fig. 20; pl. 9, fig. 16
  incertum variant, pl. 7, fig. 19
  matagordanum, pl. 8, fig. 4
  poeyanum, pl. 1, fig. 5
  subarcticum, pl. 5, fig. 13
Elphidium
  in Mississippi Sound, 139
  present-day depth range, 102
Elphidium crispum, production of, 189
Epistominella
  decorata, pl. 4, figs. 10, 11
  exigua, pl. 3, fig. 21
  vitrea, pl. 1, fig. 6; pl. 9, fig. 9
Eponides
  antillarum, pl. 1, fig. 11
  regularis, pl. 2, figs. 13, 14

tumidulus, pl. 4, figs. 4, 5
turgidus, pl. 4, figs. 6, 7
Equatorial west central fauna in Pacific, 214
Estuary
    common species, 153, pl. 8
    description of environment, 12-13
    schematic cross-section, fig. 3 (12)
Estuary and salt-marsh fauna, 153

Field methods, general, 20-30
Florida Keys, reef faunas, 183-4
Fluvial marine fauna, 159-60; pl. 9
Food, importance in distribution of faunas, 111

Galicia coast, depth distributions, 41
Glabratella wrightii, pl. 7, fig. 25a, b
Globigerina bulloides, pls. 10, 11
    distribution in Atlantic, fig. 71 (227)
    distribution in northern Gulf of Mexico, fig. 73 (231)
Globigerina eggeri, pls. 10, 11
    distribution in northern Gulf of Mexico, fig. 74 (232)
Globigerina inflata, pl. 10
Globigerina pachyderma, pl. 10
Globigerinella aequilateralis, pls. 10, 11
Globigerinita glutinata, pls. 10, 11
Globigerinoides conglobatus, pls. 10, 11
Globigerinoides ruber, pls. 10, 11
    distribution in northern Gulf of Mexico, fig. 79 (237)
Globigerinoides sacculifer, pls. 10, 11
    distribution in northern Gulf of Mexico, fig. 75 (233)
Globobulimina
    (*Desinobulimina*) *auriculata*, pl. 5, fig. 15
    mississippiensis, pl. 2, fig. 7
Globorotalia hirsuta, pl. 10
Globorotalia menardii, pls. 10, 11
    distribution in Atlantic, fig. 70 (226)
    distribution in northern Gulf of Mexico, fig. 76 (234)
    geologic range, 252
    in mixed faunas, 225
    solution effects, pl. 11

Globorotalia menardii flexuosa, geologic range, 252
Globorotalia punctulata, pl. 11
Globorotalia scitula, pl. 10
Globorotalia truncatulinoides, pls. 10, 11
    coiling direction, 247
    distribution in northern Gulf of Mexico, fig. 77 (235)
Globorotalia tumida, pls. 10, 11
    geologic range, 252
Glomospira
    charoides, pl. 4, fig. 12
    cf. *G. gordialis*, pl. 7, fig. 21
Goesella mississippiensis, pl. 2, fig. 6
Gulf of Maine
    benthonic species from "mud facies", pl. 6
    benthonic species from "sand facies", pl. 5
    density of living population, 188
    depth distribution of benthonic species, 40, 41
    depth ranges, generalized, fig. 31 (67)
    distribution of *Cibicides lobatulus*, fig. 45 (119)
    distribution of *Reophax curtus*, fig. 46 (121)
    foraminiferal evidence for sea-level change, 99
Gulf of Mexico
    beach and delta fauna, pl. 9
    benthonic species shallower than 100 m., pl. 1
    benthonic species 100-200 m., pl. 2
    benthonic species 200-500 m., pl. 3
    benthonic species deeper than 500 m., pl. 3
    benthonic species deeper than 1000 m., pl. 4
    benthonic species deeper than 2000 m., pl. 4
    continental shelf sedimentation rate, 199
    core studies of planktonic species, 244
    displaced faunas from Sigsbee Deep, fig. 40 (94)
Gulf of Paria, faunal composition, 167-171, fig. 57 (169, 170)

Gullmar Fjord, Sweden, depth boundaries, 41
Guttulina australis, pl. 7, fig. 22
Gyroidina neosoldanii, pl. 4, fig. 9
  orbicularis, pl. 3, fig. 22

Hanzawaia
  concentrica, pl. 1, fig. 10
  strattoni, pl. 7, figs. 11, 12; pl. 9, fig. 15
Haplophragmoides
  bradyi, pl. 3, fig. 23; pl. 6, fig. 2
  subinvolutum, pl. 8, figs. 17, 18
High production areas, 210, 212
Hippocrepina indivisa, pl. 6, fig. 5
Hvorslev-Stetson gravity coring tube, fig. 8 (28)
Hydrogen-ion concentration
  Baptiste Collette Subdelta, 178
  diurnal range in marshes, 177-8
  importance of, 112
  Newport Bay, California, 177-8
  San Antonio Bay, 178

Ice rafting of faunas, 267
Interdistributary bay, 160; pl. 9

Jade Bay, North Sea, faunal composition, 164-5

Kapingamarangi Atoll, coral reef fauna, 183
Kiel Bay, Germany, faunal composition, 165-6

Laboratory experiments, 104-9
  on *Streblus beccarii*, 106; figs. 42, 43 (107, 110)
Laboratory study, general, 30-38
Lagoon, common species, pl. 8
Laguna Madre
  assemblage, 133
  interpretation of faunas, 142
Laticarinina pauperata, pl. 3, fig. 25
Living Foraminifera
  collection and preservation, 24
  identification, 32-3
Living-Total ratios, benthonic species
  Mississippi Delta, fig. 59 (194)
  Mississippi Sound and Mobile Bay, fig. 58 (192)
  northern Gulf of Mexico, fig. 61 (197)
  San Antonio Bay, fig. 60 (195)
Long Island Sound
  and Block Island Sound benthonic species, fig. 48 (128)
  and Buzzards Bay Foraminifera, figs. 49, 50 (130, 131)
  attached species, 118
  benthonic faunas, 126-9
  interpretation of faunas, 145
Lower continental slope and deep-sea fauna, 259

Maracaibo Lake, faunal composition, 167
Marginal marine areas
  ecologic water masses, 138
  species composition, 153, 158
Marine marsh fauna
  common species, pl. 8
  description of environment, 11-12
  general aspects of fauna, 258
  hydrogen-ion concentration, 177-8
Marshall Islands, coral reef faunas, 181-3
Matagorda Bay
  faunas, 133
  interpretation of faunas, 143
Matsukawa-Ura Bay, Japan, faunal composition, 172-3
Mediterranean Sea
  core studies of planktonic species, 244
  depth distribution of benthonic species, 40, 41-2
Miliammina fusca, pl. 8, fig. 21a, b; pl. 9, figs. 3, 14
Miliolinella subrotunda, pl. 5, fig. 19
Mississippi Canyon, displaced faunas, 99
Mississippi Delta
  assemblage distribution, 136
  benthonic faunas, fig. 53 (137)
  density of living population, 189
  high production, 210
  interpretation of faunas, 143-5
  living-total population ratios, fig. 59 (194)
  sedimentation rate, 193-4

Mississippi Sound
   assemblage distribution, 134
   faunal distribution, fig. 52 (135)
   interpretation of fauna, 138-9, 141
   living-total population ratios, fig. 58 (192)
   sedimentation rate, 192-3

Nearshore open ocean
   common species, pl. 7
   distribution studies in Asia, 172-5
   generalized distribution of benthonic faunas, figs. 55, 56 (149-50)
   species composition of, 151-2
   turbulent zone fauna, 258-9
   water types, 8; fig. 4 (16)
Nodobaculariella cassis, pl. 1, fig. 14
Nonion
   labradoricum, pl. 5, fig. 16
   pompilioides, pl. 4, fig. 23
   tisburyensis, pl. 8, fig. 6
Nonionella
   atlantica, pl. 7, figs. 23, 24
   opima, pl. 1, fig. 12; pl. 7, figs. 26, 27; pl. 9, figs. 6, 7
North Pacific
   composition of planktonic faunas, fig. 68 (215)
   generalized distribution of planktonic faunas, fig. 69 (216)
North Sea, faunal composition in Jade Bay, 164-5
Northern Gulf of Mexico
   change of sea-level, 101-2
   comparison of faunas and sediments, 114-7; fig. 44 (116)
   density of living population, 188-9
   depth distribution of benthonic forms, 40
   depth ranges of abundant species, figs. 14-17 (48-51)
   depth ranges of planktonic species, 241
   displaced faunas in cores from Sigsbee Deep, 92-6
   distribution of *Globigerina bulloides*, fig. 73, (231)
   distribution of *Globigerina eggeri*, fig. 74 (232)
   distribution of *Globigerinoides ruber*, fig. 79 (237)
   distribution of *Globigerinoides sacculifer*, fig. 75 (233)
   distribution of *Globorotalia menardii*, fig. 76 (234)
   distribution of *Globorotalia truncatulinoides*, fig. 77 (235)
   distribution of *Pulleniatina obliquiloculata*, fig. 78 (236)
   generalized depth ranges, continental shelf, 53; figs. 18-21 (54-57)
   generalized depth ranges of species, figs. 22-28 (58-64)
   inner continental shelf species, 80
   interpretation of planktonic faunas in cores, fig. 83 (251)
   living-total population benthonic ratios, fig. 61 (197)
   outer continental shelf species, 80
   planktonic distribution, 216-7, 228-239
   planktonic forms on continental shelf, fig. 80 (240)
   planktonic-benthonic ratios, 242
   post-glacial sedimentation, 247, 249
   sedimentation rate, 191-201
   size of planktonic population 216-7
   species restricted to 90-100 m., 80
   thickness of Recent planktonic fauna, 247,249
   turbulent zone species, 80
Nouria polymorphinoides, pl. 1, fig. 13

Ojo de Liebre Lagoon, 175
   high production, 210
Open-ocean species, common nearshore, pl. 7
Orange-peel dredge, fig. 7 (25)
Orbulina universa, pl. 10
   living depth, 218-20
Osangularia cultur, pl. 3, fig. 24

Pacific
   displaced faunas, 98
   plankton faunas, 214-5
Palmerinella gardenislandensis, pl. 9, figs. 2, 12
Peper Bay, Java, faunal composition, 174-5
Permanent thermocline, 5
Persian Gulf
   evidence for lowered sea-level, 102
   sedimentation rate, 201

Plankton tows, 27, 30
Planulina
  ariminensis, pl. 3, figs. 7, 8
  exorna, pl. 1, fig. 15
  foveolata, pl. 2, figs. 9, 10
Plectina apicularis, pl. 4, fig. 1
Pleistocene
  correlation in cores, 246-7, 249, 252
  displacement of Counter Current, 245-6
Pontigulasia compressa, pl. 8, fig. 22
Poponessett Bay, Mass.
  marsh study, 178-9
  seasonal production, 205
Pressure effect on organisms, 123
Proteonina atlantica, pl. 7, fig. 29
Protoplasm, recognition, 32
Pseudoclavulina mexicana, pl. 2, fig. 20
Pullenia bulloides pl. 4, fig. 14
Pulleniatina obliquiloculata, pls. 10, 11
  distribution in northern Gulf of Mexico, fig. 78 (236)

Quantitative studies, probable error, 33-35; fig. 12 (34)
Quinqueloculina
  arctica, pl. 5, fig. 18
  compta, pl. 1, fig. 16; pl. 7, figs. 30, 31; pl. 9, figs. 19,20
  horrida, pl. 1, fig. 17
  seminulum, pl. 5, fig. 17; pl. 9, figs. 21,22
  venusta, pl. 4, fig. 24
Quinqueloculina poeyana, size histogram of living in San Antonio Bay, fig. 66 (209)

Rectobolivina advena, pl. 1, fig. 20
Recurvoides
  turbinatus, pl. 6, figs. 6, 7
  sp., pl. 8, figs. 24, 25
Red Sea, depth boundaries, 41
Reophax
  curtus, pl. 6, figs. 12,13
  curtus variant, pl. 7, fig. 32
  dentaliniformis, pl. 8, fig. 8
  nanus, pl. 8, fig. 3
  scottii, pl. 6, fig. 14; pl. 7, fig. 33
Reophax curtus, distribution in Gulf of Maine, fig. 46 (121)

Reproduction rate, experiments with *Streblus beccarii,* 206
Residual faunas
  San Diego area, 202, 204
  Todos Santos Bay, 204
Reussella atlantica, pl. 2, fig. 21; pl. 7, fig. 34
Rhone Delta
  depth boundaries, 41
  faunal composition, 161, 164
Rosalina
  bertheloti, pl. 2, figs. 11,12
  columbiensis, pl. 5, fig. 9a, b; pl. 7, fig. 13a, b
  floridana, pl. 7, figs. 14,15
Rose Bengal staining technique, 32
"*Rotalia*"
  rolshauseni, pl. 7, fig. 35a, b
  translucens, pl. 3, figs. 13, 14
Runoff and faunal patterns, 146
Russian Seas, depth distribution of benthonic species, 40, 42-3

Salinity
  as affecting distributions, 108-9
  experimental effects on *Streblus beccarii,* 109, fig. 43 (110)
  importance in distinguishing ecologic water masses, 109
Salton Sea, Calif., abnormality in specimens, 271
Sample preparation, 37-8
Sample splitters, 33
Sampling gear
  Clarke-Bumpus plankton sampler, 27, 30; fig. 10 (31)
  coring tube, fig. 5 (22)
  cutting device, fig. 6 (25)
  Hvorslev-Stetson coring tube, 27; fig. 8 (28)
  orange-peel dredge, 24-5; fig. 7 (25)
  piston coring tube, Kullenberg, 27, fig. 9 (29)
  plankton tow net, fig. 11 (31)
  vanVeen grab, 26
Sampling methods, general 21-7
San Antonio Bay
  assemblage distribution, 129, 133
  distribution of faunas, fig. 51 (132)
  high production, 210

histograms of living *Streblus beccarii*, fig. 65 (208)
interpretation of faunas, 142
living-total population ratios, fig. 60 (195)
seasonal production, 206, 209
sedimentation rate, 195-6
size histograms of living *Ammotium salsum*, fig. 64 (207)
size histograms of living *Quinqueloculina poeyana*, fig. 66 (209)
Thecamoebina, 180
San Diego, Calif.
  depth boundaries, 42
  high production, 212
  residual sediments, 202, 204
  total populations, fig. 62 (203)
San Diego Trough
  displaced forms in cores, figs. 37-39 (91, 93)
  displaced sediments, 98-9
  evidence for displacement, 90-92
Santa Monica Bay, Calif., seasonal production, 205
Sea-level change
  Baja California, 102
  evidence, 99-102
  northern Gulf of Mexico, 101-2
  Persian Gulf, 102
Seasonal layer
  effect on depth distributions, 120, 122
  nature of on continental shelf, 17
Sedimentation rate
  Baja California, 201
  Bay of Bengal, 201
  discussion, 199-204
  Mississippi Delta, 193-4
  Mississippi Sound, 191-3
  northern Gulf of Mexico, 196-201
  Persian Gulf, 201
  procedures for determining relative, 190-1
  San Antonio Bay, 195-6
Shark Bay, Western Australia, 175
Sigmoilina
  distorta, pl. 2, fig. 15
  schlumbergeri, pl. 4, fig. 19
Sigsbee Deep, displaced faunas, 92-6; fig. 40 (94)
Siphonina
  bradyana, pl. 3, fig. 12

pulchra, pl. 2, figs. 18, 19
Siphotextularia
  curta, pl. 4, fig. 15
  rolshauseni, pl. 4, fig. 16
Size of samples, measure of, 36-7
Skagerak, depth boundaries, 41
South America, nearshore distribution studies, 167-8, 171-2
Spiroplectammina biformis, pl. 6, figs. 15, 16
Staining technique, 32
Standing crop of benthonic Foraminifera, 188
Staten Island, New York, marshes, 178
Streblus beccarii and variants, pl. 1, figs. 18, 19; pl. 7, fig. 28; pl. 9, figs. 5, 23
  as affected by salinity, 109, fig. 43 (110)
  as affected by temperature, 106-7; fig. 42 (107)
  in Mississippi Sound, 139
  living histograms in San Antonio Bay, fig. 65 (208)
  present-day depth range, 102
  reproduction rate, 206
Subarctic fauna in Pacific, 214
Submarine canyons, as related to displaced sediments, 98, 99
Substrate
  attached species, Long Island Sound, 118
  correlation with faunas, 113-4
  effects on distribution, 113-8
Surface circulation, 8, 9
Temperature
  affecting distributions, 103-8
  experiments on *Streblus beccarii*, 106-7; fig. 42 (107)
  general seasonal gradients in mid-latitudes and tropics, fig. 2 (7)
  mean monthly sea surface in Atlantic, fig. 1 (6)
  range in lagoons, 15
  related to depth faunas, fig. 41 (105)
  relation to planktonic Foraminifera distribution, 224
  seasonal layer, 5; fig. 2a (7)
  variation in North Atlantic, 4-5

INDEX 297

Textularia
 earlandi, pl. 8, fig. 10
 foliacea occidentalis, pl. 2, fig. 22
 mayori, pl. 1, fig. 22; pl. 7, fig. 36
 torquata, pl. 6, figs. 17, 18
Thecamoebina
 common species, pl. 8, (156)
 distribution studies, 179-81
 in San Antonio Bay, 180
Tiphotrocha comprimata, pl. 8, figs. 26, 27
Todos Santos Bay
 depth distribution of living populations, fig. 67 (211)
 depth of faunas, 82-90
 high production, 210
 living species depth ranges, figs. 32-36 (83-87)
 residual faunas, 204
 seasonal abundance of average living populations, fig. 63 (205)
 seasonal production, 204-5
Tokyo Bay, Japan, faunal composition, 173
Total populations
 estimate of, 33-35
 in non-depositional areas, 202
 study, 33-37

Trifarina bradyi, pl. 3, fig. 10
Trinidad, Thecamoebina study, 179-80
Trochammina
 advena, pl. 6, fig. 19a, b
 compacta, pl. 8, fig. 9a, b
 globulosa, pl. 4, fig. 13
 inflata, pl. 5, figs. 20, 21; pl. 8, fig. 28a, b
 lobata, pl. 5, fig. 22a, b
 macrescens, pl. 8, fig. 29a, b
 quadriloba, pl. 6, fig. 20a, b
 squamata, pl. 6, figs. 21, 22
Turbulent zone species, 80
Tyrrhenian Sea, core studies, 244

Uvigerina
 auberiana, pl. 4, fig. 18
 flintii, pl. 2, fig. 16
 hispido-costata, pl. 3, fig. 16
 laevis, pl. 2, fig. 24
 parvula, pl. 2, fig. 23

Virgulina
 advena, pl. 4, fig. 20
 complanata, pl. 5, fig. 23
 fusiformis, pl. 5, fig. 24
 pontoni, pl. 7, fig. 37
 punctata, pl. 1, fig. 21
 tessellata, pl. 3, fig. 15

Wadden Sea, Netherlands, faunal composition, 166-7